AP

CALCULUS AB

Study Guide and Full Practice Exam

Cover by Mirvoxid Press
Illustrations by Mirvoxid Press

Content:

Key Exam Details

The AP® Calculus AB exam is a 3-hour and 15-minute, end-of-course test that consists of 45 multiple-choice questions (50% of the exam) and 6 free-response questions (50% of the exam).

The exam will include questions on the various topics covered in the book:

- **Limits and Continuity:** 10–12% of test questions
- **Differentiation: Definition and Basic Derivative Rules:** 10–12% of test questions
- **Differentiation: Composite, Implicit, and Inverse Functions:** 9–13% of test questions
- **Contextual Applications of Differentiation:** 10–15% of test questions
- **Applying Derivatives to Analyze Functions:** 15–18% of test questions
- **Integration and Accumulation of Change:** 17–20% of test questions
- **Differential Equations:** 6–12% of test questions
- **Applications of Integration:** 10–15% of test questions

This guide offers an overview of the core tested subjects, along with sample AP multiple-choice questions that are like the questions you'll see on test day.

Limits and Continuity

Around 10–12% of the questions on your AP Calculus AB exam will feature **Limits and Continuity** questions.

Limits

The limit of a function f as x approaches c is L if the value of f can be made arbitrarily close to L by taking x sufficiently close to c (but not equal to c). If such a value exists, this is denoted $\lim_{x \to c} f(x) = L$. If no such value exists, we say that the limit does not exist, abbreviated DNE.

There are various methods to determine limits, including the use of tables, graphs, and algebraic techniques.

Here's an example:

The table below provides several values of a function.

x	0.9	0.99	0.999	1.001	1.01	1.1
$f(x)$	2.488	2.490	2.499	2.501	2.504	2.513

Based on these values, it appears that $\lim_{x\to 1} f(x) = 2.5$, since the values of the function are growing close to 2.5 as c approaches 1.

Crucial algebraic techniques for determining limits involve the utilization of factoring and rationalizing radical expressions. Additional useful tools are provided by the following properties.

Suppose $\lim_{x\to c} f(x) = L$, $\lim_{x\to c} g(x) = M$, $\lim_{x\to L} h(x) = N$, and a is any real number. Then,

- $\lim_{x\to c}[f(x)+g(x)] = L+M$
- $\lim_{x\to c}[f(x)-g(x)] = L-M$
- $\lim_{x\to c}[af(x)] = aL$
- $\lim_{x\to c}\frac{f(x)}{g(x)} = \frac{L}{M}$, as long as $M \neq 0$
- $\lim_{x\to c} h(f(x)) = N$

When it comes to evaluating limits for common functions, all you need to do is evaluate the function at the given point c, as long as the function is defined at that point. These functions encompass a variety of mathematical concepts, such as polynomials, rational expressions, exponentials, logarithms, and trigonometry.

Two special limits that are important in calculus are $\lim_{x\to 0}\frac{\sin x}{x} = 1$ and $\lim_{x\to 0}\frac{1-\cos x}{x} = 0$.

One-Sided Limits

Sometimes we are interested in the value that a function f approaches as x approaches c from only a single direction. If the values of f get arbitrarily close to L as x approaches c while taking on values greater than c, we say $\lim_{x\to c^+} f(x) = L$. Similarly, if x is taking on values less than c, we write $\lim_{x\to c^-} f(x) = L$.

We can now characterize limits by saying that $\lim_{x\to c} f(x)$ exists if and only if both $\lim_{x\to c^+} f(x)$ and $\lim_{x\to c^-} f(x)$ exist and have the same value. A limit, then, can fail to exist in a few ways:

- $\lim_{x\to c^+} f(x)$ does not exist
- $\lim_{x\to c^-} f(x)$ does not exist
- Both of these one-sided limits exist, but have different values

Example

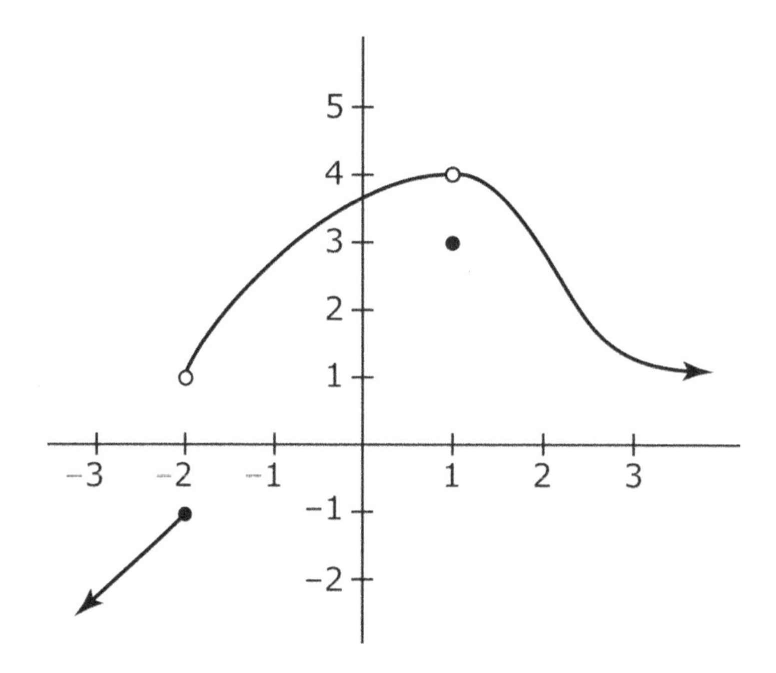

The function shown has the following limits:

- $\lim_{x \to -2^-} f(x) = -1$
- $\lim_{x \to -2^+} f(x) = 1$
- $\lim_{x \to -2} f(x)$ DNE
- $\lim_{x \to 1^-} f(x) = 4$
- $\lim_{x \to 1^+} f(x) = 4$
- $\lim_{x \to 1} f(x) = 4$

Note that $f(1) = 3$, but this is irrelevant to the value of the limit.

Infinite Limits, Limits at Infinity, and Asymptotes

When a function has a vertical asymptote at $x = c$, the behavior of the function can be described using infinite limits. If the function values increase as they approach the asymptote, we say the limit is $\times$, whereas if the values decrease as they approach the asymptote, the limit is $-\times$. It is important to realize that these limits do not exist in the same sense that we described earlier; rather, saying that a limit is $\pm\times$ is simply a convenient way to describe the behavior of the function approaching the point.

We can also extend limits by considering how the function behaves as $x \to \pm\infty$. If such a limit exists, it means that the function approaches a horizontal line as x increases or decreases without bound. In other words, if $\lim_{x \to \pm\infty} f(x) = L$, then f has a horizontal asymptote $y = L$. It is possible for a function to have two horizontal asymptotes, since it can have different limits as $x \to \infty$ and $x \to -\infty$.

Example

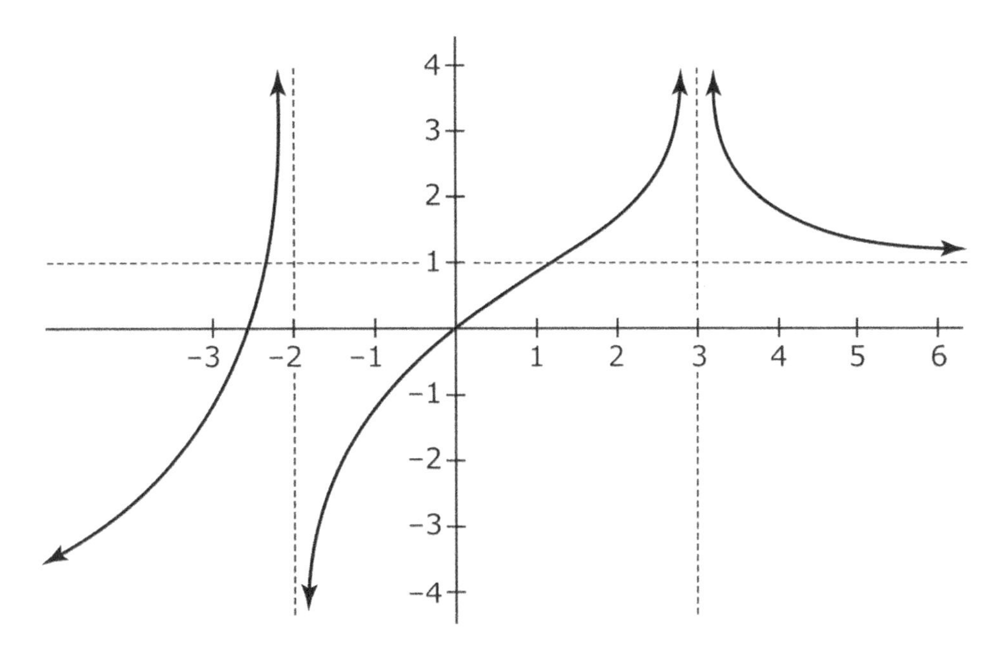

The function above has vertical asymptotes at $x = -2$ and $x = 3$, and a horizontal asymptote at $y = 1$. Looking at the graph, we can determine the following limits:

- $\lim_{x \to -2^-} f(x) = \infty$
- $\lim_{x \to -2^+} f(x) = -\infty$
- $\lim_{x \to 3} f(x) = \infty$
- $\lim_{x \to \infty} f(x) = 1$

The Squeeze Theorem

The Squeeze Theorem states that if the graph of a function lies between the graphs of two other functions, and if the two other functions share a limit at a certain point, then the function in between also shares that same limit. More formally, if $f(x) \leq g(x) \leq h(x)$ for all x in some interval containing c, and if $\lim_{x \to c} f(x) = \lim_{x \to c} h(x) = L$, then $\lim_{x \to c} g(x) = L$ as well.

Example

The sine function satisfies $-1 \le \sin x \le 1$ for all real numbers *x*, so $-1 \le \sin\left(\frac{1}{x}\right) \le 1$ is also true for all real numbers *x*. Multiplying this inequality by x^2, we obtain $-x^2 \le x^2 \sin\left(\frac{1}{x}\right) \le x^2$. Now the functions on the left and right of the inequality, x^2 and $-x^2$, both have limits of 0 as $x \to 0$. Therefore, we can conclude that $\lim_{x \to 0} x^2 \sin\left(\frac{1}{x}\right) = 0$ also.

Continuity

The function *f* is said to be continuous at the point $x = c$ if it meets the following criteria:

1. $f(c)$ exists
2. $\lim_{x \to c} f(x)$ exists
3. $\lim_{x \to c} f(x) = f(c)$

In other words, the function must have a limit at *c*, and the limit must be the actual value of the function.

Each of the previously mentioned criteria can fail, resulting in a discontinuity at at $x = c$.

Consider the following three graphs:

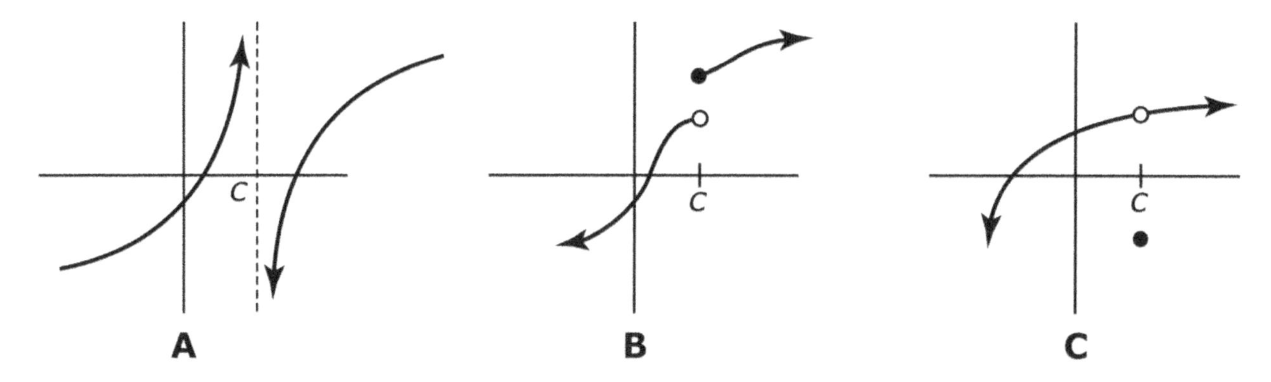

In graph A, the function is not defined at *c*. In graph B, the function is defined at *c*, but the limit as $x \to c$ does not exist due to the one-sided limits being different. In graph C, the function is defined at *c* and the limit as $x \to c$ exists, but they are not equal to each other.

The discontinuity in graph B is referred to as a jump discontinuity, since it is caused by the graph jumping when it reaches at $x = c$. In contrast to this is the situation in graph C, where the discontinuity could be fixed by moving a single point; it occurs whenever the second condition above is satisfied and is called a removable discontinuity. If $\lim f(x)$ exists, but f has a discontinuity at $x = c$ because it fails one of the other conditions, the discontinuity can be removed by defining or redefining $f(c)$ to be equal to the limit at that point.

A function is considered continuous on an interval when it maintains continuity at every point within that interval. Here are some types of functions that are continuous at every point in their domains:

- Polynomial
- Rational
- Power
- Exponential
- Logarithmic
- Trigonometric

If f is a function with different parts defined in different ways, we can determine if it is continuous by examining its behaviour at the points where the different parts meet. For continuity at a boundary point, it is necessary for the functions on both sides of the point to yield identical results when evaluated at the point

Example

Consider the function $f(x) = \begin{cases} 3x+2 & x \le 0 \\ x^2 - 1 & 0 < x < 4 \\ 5 + 10\sin\dfrac{\pi x}{8} & x \ge 4 \end{cases}$

Each of the component functions are continuous at all real numbers, so we need only check continuity at $x = 0$ and $x = 4$. For $x = 0$, the function to the left is $3(0)+2=2$, and to the right we have $(0)^2 - 1 = -1$. These are not equal, so there is a jump discontinuity at $x = 0$.

Looking now at $x = 4$, the results from the functions on the two sides are $4^2 - 1 = 15$ and $5 + 10\sin\dfrac{4\pi}{8} = 15$. Since these are equal, the function is continuous at $x = 4$.

Intermediate Value Theorem

The Intermediate Value Theorem applies to continuous functions on an interval $[a,b]$. If d is any value between $f(a)$ and $f(b)$, then there must be at least one number c between a and b such that $f(c) = d$.

Example

Consider $f(x) = e^x - 2$, which is continuous everywhere. We have $f(0) = e^0 - 2 = -1$, and $f(1) = e - 2$, which is certainly positive. If we take $d = 0$ in the statement of the theorem, then d is between $f(0)$ and $f(1)$. Therefore, the Intermediate Value Theorem guarantees at least one value c between 0 and 1 with the property that $f(c) = 0$. This value, of course, is $c = \ln 2$.

Practice Limits and Continuity Questions

Do not use a calculator for the following two problems.

Suppose the graph of $F(x)$ is given by the following:

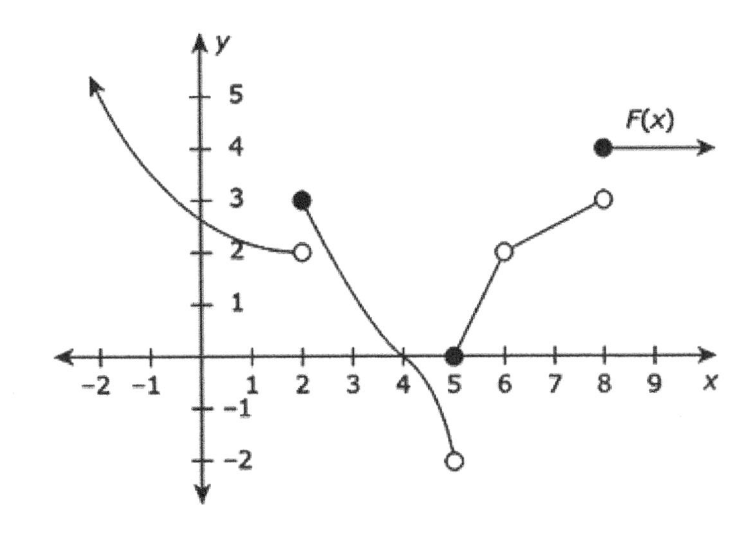

Which of the following statements is TRUE?

A. $\lim_{x \to 5^+} F(x) = -2$

B. $\lim_{x \to 8^-} F(x) = 3$

C. $\lim_{x \to 6} F(x)$ does not exist.

D. $\lim_{x \to 2} F(x) = 3$

The correct answer is B. This is true since the closer you take x values from the left side of 8, the closer the corresponding y-values on the graph of $F(x)$ get to 3. Choice A is actually the value of the left-sided limit at 5. The right-sided limit at 5 is 0. The limit in choice C is actually equal to 6. Remember, a function need not be defined at an x-value in order to have a limit there. Choice D is incorrect because even though $F(2) = 3$, the y-values get close to 2, not 3.

Compute the limit: $\lim_{x \to \pi} \left[\frac{\sin(\pi - x)}{x - \pi} + \frac{|-x|}{2\pi} - 3 \right]$.

A. $-\frac{9}{2}$

B. $-\frac{7}{2}$

C. $-\frac{5}{2}$

D. $-\frac{3}{2}$

The correct answer is B. Use the limit theorem "limit of a sum/difference is the sum/difference of the limits":

$$\lim_{x\to\pi}\left[\frac{\sin(\pi-x)}{x-\pi}+\frac{|-x|}{2\pi}-3\right]=\lim_{x\to\pi}\frac{\sin(\pi-x)}{x-\pi}+\lim_{x\to\pi}\frac{|-x|}{2\pi}-\lim_{x\to\pi}3$$

Now, compute each limit separately:

$$\lim_{x\to\pi}\frac{\sin(\pi-x)}{x-\pi}=\lim_{x\to\pi}\frac{\sin(\pi-x)}{-(\pi-x)}=-\lim_{x\to\pi}\frac{\sin(\pi-x)}{(\pi-x)}=-(1)=-1$$

$$\lim_{x\to\pi}\frac{|-x|}{2\pi}=\frac{|-\pi|}{2\pi}=\frac{\pi}{2\pi}=\frac{1}{2}\text{ (by continuity)}$$

$$\lim_{x\to\pi}3=3$$

Substituting these into the above equation yields

$$\lim_{x\to\pi}\left[\frac{\sin(\pi-x)}{x-\pi}+\frac{|-x|}{2\pi}-3\right]=\lim_{x\to\pi}\frac{\sin(\pi-x)}{x-\pi}+\lim_{x\to\pi}\frac{|-x|}{2\pi}-\lim_{x\to\pi}3=-1+\frac{1}{2}-3=-\frac{7}{2}.$$

You may use a graphing calculator to solve the following problem.

Consider the function $H(x)=\begin{cases}x\cos\left(\frac{1}{x}\right), & x\neq 0\\ 3, & x=0\end{cases}$. Which of the following statements, if either, is true?

(I) $H(x)$ appears to have a removable discontinuity at $x = 0$.
(II) $H'(0)\approx 0$.

A. I only
B. II only
C. Both I and II
D. Neither I nor II

The correct answer is A. Use the graphing calculator to graph $H(x)$:

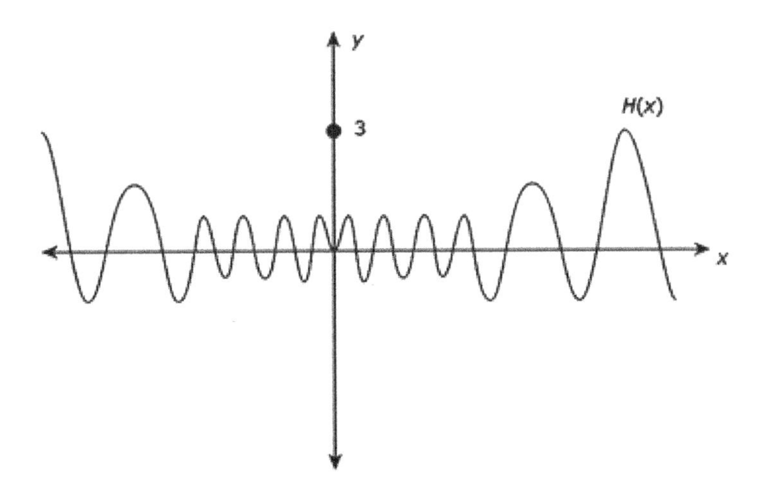

If you continue to zoom in on the origin, it becomes evident that $\lim_{x \to 0} H(x) = 0$. Since $H(0) = 3$, there is a removable discontinuity at $x = 0$. So, (I) is true. As such, (II) must be false because $H(x)$ is discontinuous at $x = 0$, and hence cannot be differentiable at $x = 0$.

Differentiation: Definition and Fundamental Properties

On your AP exam, 10–12% of questions will cover Differentiation: Definition and Fundamental Properties.

Definition of the Derivative

The average rate of change of a function *f* over the interval from $x = a$ to $x = a + h$ is $\frac{f(a+h)-f(a)}{h}$. Alternatively, if $x = a + h$, this can be written as $\frac{f(x)-f(a)}{x-a}$. When h is made smaller, so that it approaches 0, the limit that results is called the instantaneous rate of change of *f* at $x = a$, or the derivative of *f* at $x = a$, and is denoted $f'(a)$.

That is, $f'(a) = \lim_{h \to 0} \frac{f(a+h)-f(a)}{h}$, or equivalently, $f'(a) = \lim_{x \to a} \frac{f(x)-f(a)}{x-a}$.

If this limit exists, *f* is said to be differentiable at *a*. Graphically, $f'(a)$ represents the slope of line tangent to the graph of $f(x)$ at the point where $x = a$. Therefore, the line tangent to $f(x)$ at $x = a$ is $y - f(a) = f'(a)(x - a)$.

If the function $y = f(x)$ is differentiable at all points in some interval, we can define a new function on that interval by finding the derivative at every point. This new function, called the derivative of *f*, can be denoted $f'(x)$, y', or $\frac{dy}{dx}$, and is defined by $f'(x) = \lim_{h \to 0} \frac{f(x+h)-f(x)}{h}$.

The value of the derivative at a particular point, $x = a$, can then be denoted $f'(a)$ or $\left.\frac{dy}{dx}\right|_{x=a}$.

If *f* is differentiable at $x = a$, then it also must be continuous at $x = a$. In other words, if a function fails to be continuous at a point, it cannot possibly be differentiable at that point. Another way that differentiability can fail is via the presence of sharp turns or cusps in a graph.

Example

The graph of a function is shown below.

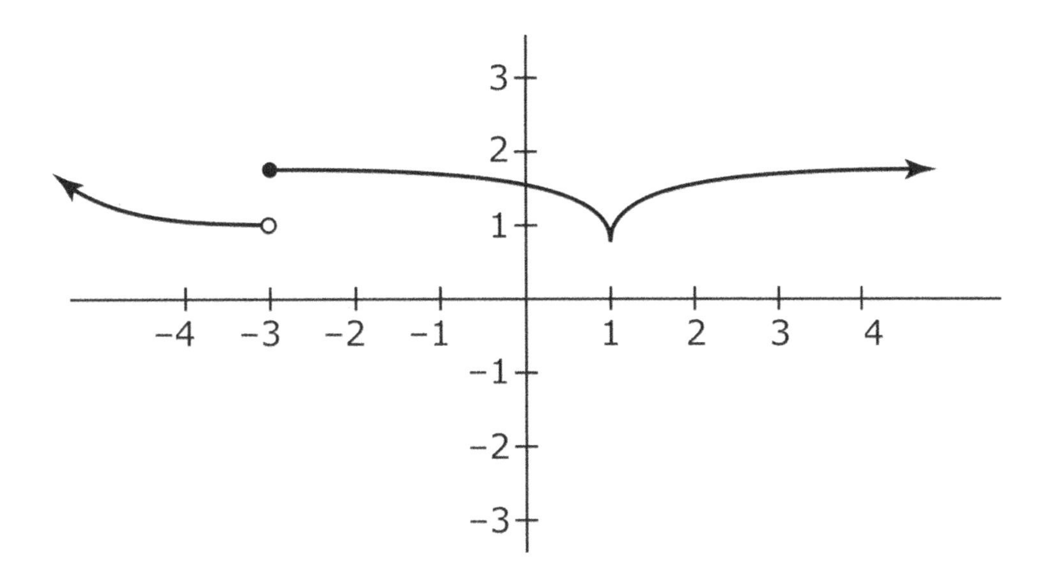

This function is differentiable everywhere except $x = -3$ (because it is not continuous there) and $x = 1$ (because it has a cusp there).

> **Free Response Tip**
>
> When you have specific function values, you can approximate the derivative at a point by calculating the average rate of change between the surrounding points. For instance, if you have the function's values at x = 3, 4, and 5, you can estimate the derivative at 4 by calculating the average rate of change between 3 and 5.

Basic Derivatives and Rules

There are several rules that can be used to find derivatives. Assume *f* and *g* are differentiable functions, and *c* is a real number.

- The constant rule: $\frac{d}{dx}c = 0$
- The power rule: $\frac{d}{dx}x^n = nx^{n-1}$, for any real number *n*

- The sum rule: $\frac{d}{dx}[f(x)+g(x)]=f'(x)+g'(x)$
- The difference rule: $\frac{d}{dx}[f(x)-g(x)]=f'(x)-g'(x)$
- The constant multiple rule: $\frac{d}{dx}[cf(x)]=cf'(x)$
- The product rule: $\frac{d}{dx}[f(x)g(x)]=f'(x)g(x)+f(x)g'(x)$
- The quotient rule: $\frac{d}{dx}\left[\frac{f(x)}{g(x)}\right]=\frac{f'(x)g(x)-f(x)g'(x)}{[g(x)]^2}$

As special cases of the power rule, note that $\frac{d}{dx}(cx)=c$, and $\frac{d}{dx}x=1$.

In addition to these rules, the derivatives of some common functions are as follows:

$f(x)$	$f'(x)$
e^x	e^x
$\ln x$	$\frac{1}{x}$
$\sin x$	$\cos x$
$\cos x$	$-\sin x$
$\tan x$	$\sec^2 x$
$\sec x$	$\sec x \tan x$
$\csc x$	$-\csc x \cot x$
$\cot x$	$-\csc^2 x$

The last four of these can be derived using the product or quotient rule along with the derivatives of $\sin x$ and $\cos x$.

Example

If $f(x)=3x^2\sin x$, then using the power and product rules, we have $f'(x)=6x\sin x+3x^2\cos x$.

Example

Let $y=\frac{2xe^x}{\cos x}$. Using the product rule, the derivative of the numerator is $2e^x+2xe^x$. Therefore, by the quotient rule, the derivative of the entire function is $y=\frac{\left(2e^x+2xe^x\right)\cos x-2xe^x\left(-\sin x\right)}{\cos^2 x}$.

Practice Differentiation: Definition and Fundamental Properties Questions

Do not use a calculator for the following problems.

Find the equation of the tangent line to the curve $y^3 + yx^2 + x^2 - 3y^2 = 0$ at the point $(-1,1)$.

A. $y = 4x + 5$
B. $y = 0$
C. $y = -2x - 1$
D. $y = -\frac{2}{5}x + \frac{3}{5}$

The correct answer is C. First, implicitly differentiate both sides of the equation with respect to x and solve for y' :

$$3y^2y' + y \cdot 2x + x^2 \cdot y' + 2x - 6yy' = 0$$

$$y'\left(3y^2 + x^2 - 6y\right) = -2x - 2xy$$

$$y' = \frac{-2x - 2xy}{3y^2 + x^2 - 6y}$$

The slope of the tangent line is $y'\big|_{x=-1,\,y=1}$, computed as follows:

$$y'\big|_{x=-1,\,y=1} = \frac{-2(-1) - 2(-1)(1)}{3(1)^2 + (-1)^2 - 6(1)} = -2.$$

So, the equation of the tangent line to this curve at the point $(-1,1)$ is $y - 1 = -2(x + 1)$, or equivalently, $y = -2x - 1$.

If $y(x) = \dfrac{x}{x^{-2} + x^{-3}}$, compute $y'(1)$.

A. $-\frac{7}{4}$
B. $-\frac{3}{4}$
C. $-\frac{1}{5}$
D. $\frac{7}{4}$

The correct answer is D. Use the quotient rule:

$$y'(x) = \frac{\left(x^{-2}+x^{-3}\right)\cdot \frac{d}{dx}x - x\cdot \frac{d}{dx}\left(x^{-2}+x^{-3}\right)}{\left(x^{-2}+x^{-3}\right)^2} = \frac{\left(x^{-2}+x^{-3}\right)\cdot 1 - x\cdot\left(-2x^{-3}-3x^{-4}\right)}{\left(x^{-2}+x^{-3}\right)^2}$$

Now, simply substitute $x = 1$ in to obtain

$$y'(1) = \frac{\left(1^{-2}+1^{-3}\right)\cdot 1 - 1\cdot\left(-2\cdot 1^{-3}-3\cdot 1^{-4}\right)}{\left(1^{-2}+1^{-3}\right)^2} = \frac{2-(-5)}{2^2} = \frac{7}{4}$$

What is the equation of the line passing through the point (–2, 1) that is perpendicular to the tangent line to the curve $y(x) = \sec(x) + 2\sin(x)$ at $x = \pi$?

A. $y = -\frac{1}{2}x$
B. $y = \frac{1}{2}x + 2$
C. $y = -2x - 3$
D. $y = -2x + 2\pi - 1$

The correct answer is B. We already have a point on the line, namely (–2, 1). It remains to determine its slope. Since the line must be perpendicular to the tangent line, its slope is the negative reciprocal of the slope of the tangent line at $x = \pi$; that is, its slope equals $-\frac{1}{y'(\pi)}$. Observe that

$$y'(x) = \sec x \tan x + 2\cos x$$
$$y'(\pi) = \sec \pi \tan \pi + 2\cos \pi = (-1)(0) + 2(-1) = -2$$

Thus, the slope of the line we seek is $\frac{1}{2}$. Using point-slope formula, we find that the equation of the desired line is $y - 1 = \frac{1}{2}(x+2)$, or equivalently $y = \frac{1}{2}x + 2$.

Differentiation: Composite, Implicit, and Inverse Functions

A significant portion of the questions on your AP Calculus exam will focus on Differentiation: Composite, Implicit, and Inverse Functions, comprising approximately 9–13% of the total.

Chain Rule

The chain rule makes it possible to differentiate composite functions. If $y = f(g(x))$, then the chain rule states that $y' = f'(g(x)) \cdot g'(x)$. In alternative notation, if $y = f(u)$ and $u = g(x)$, then $\frac{dy}{dx} = \frac{dy}{du} \cdot \frac{du}{dx}$.

Example

If $y = \sin(6x^2 - 5x)$, then $y = f(g(x))$, where $f(x) = \sin x$ and $g(x) = 6x^2 - 5x$. Since $f'(x) = \cos x$ and $g'(x) = 12x - 5$, the chain rule gives us $y' = \cos(6x^2 - 5x) \cdot (12x - 5)$.

The chain rule can be extended to compositions of more than two functions by considering that $g(x)$, as described previously, may itself be a composition. If $y = f(g(h(x)))$, two applications of the chain rule yield $y' = f'(g(h(x))) \cdot g'(h(x)) \cdot h'(x)$.

Example

Suppose $f(x) = 3\tan^2\left(\frac{x}{3}\right)$. This is a composition of the functions $3x^2$, $\tan x$, and $\frac{x}{3}$. Its derivative is $f'(x) = 6\tan\left(\frac{x}{3}\right) \cdot \sec^2\left(\frac{x}{3}\right) \cdot \frac{1}{3}$.

Implicit Differentiation and Inverse Functions

A function may sometimes be presented in implicit, rather than explicit, form. That is, it may not be given as $y = f(x)$, but rather as an equation that relates x and y to each other. In such cases, we say that y is implicitly defined as a function of x. *Implicit differentiation* is the process of finding the derivative $\frac{dy}{dx}$ for such functions, and it is accomplished by applying the chain rule.

Example

Consider the equation $y^3 + x^3 + xy = 5$. Differentiating both sides of the equation with respect to *x*, and remembering that we are assuming that *y* is, in fact, a function of *x* (so that the chain rule applies), we get the following:

$$\frac{d}{dx}\left(y^3 + x^3 + xy = 5\right) = \frac{d}{dx}(5)$$
$$3y^2 \cdot \frac{dy}{dx} + 3x^2 + 1 \cdot y + x \cdot 1 \cdot \frac{dy}{dx} = 0$$

Note that differentiating xy required an application of the product rule, and that every time an expression in terms of *y* was differentiated, the derivative was multiplied by $\frac{dy}{dx}$. Now all of the terms with $\frac{dy}{dx}$ can be gathered on one side of the equation, and $\frac{dy}{dx}$ can be solved for:

$$3y^2 \cdot \frac{dy}{dx} + x \cdot \frac{dy}{dx} = -y - 3x^2$$
$$\frac{dy}{dx}\left(3y^2 + x\right) = -y - 3x^2$$
$$\frac{dy}{dx} = \frac{-y - 3x^2}{3y^2 + x}$$

This technique can also be applied to find the derivatives of inverse functions. Consider an invertible function *f*, with inverse f^{-1}. By definition, this means that $f\left(f^{-1}(x)\right) = x$. Now, differentiating both sides with respect to *x*, we get $f'\left(f^{-1}(x)\right) \cdot \left(f^{-1}\right)'(x) = 1$. Solving for $\left(f^{-1}\right)'(x)$, we have $\left(f^{-1}\right)'(x) = \frac{1}{f'\left(f^{-1}(x)\right)}$.

Example

If $f(3) = 5$, and $f'(3) = 2$, then $\left(f^{-1}\right)'(5) = \frac{1}{f'(f^{-1}(5))} = \frac{1}{f'(3)} = \frac{1}{2}$.

Applying this rule to the inverse trigonometric functions, we can find the following derivatives:

$\arcsin x$	$\frac{1}{\sqrt{1-x^2}}$
$\arccos x$	$\frac{-1}{\sqrt{1-x^2}}$
$\arctan x$	$\frac{1}{1+x^2}$
$\text{arccot}\, x$	$\frac{-1}{1+x^2}$
$\text{arcsec}\, x$	$\frac{1}{x\sqrt{x^2-1}}$
$\text{arccsc}\, x$	$\frac{-1}{x\sqrt{x^2-1}}$

Higher Order Derivatives

The derivative f' of a function f is itself a function that may be differentiable. If it is, then its derivative is f'', called the *second derivative* of f. The relationship of f' and f'' is identical to the relationship between f and f'. Similarly, the derivative of f'' is f''', the third derivative of f. This process can continue indefinitely, as long as the functions obtained continue to be differentiable. After three, the notation changes, so that the 4th derivative of f is denoted $f^{(4)}$, and the nth derivative is $f^{(n)}$.

If $y = f(x)$, then higher order derivatives are also denoted $y'', y''', y^{(4)}, \ldots, y^{(n)}, \ldots$, or
$\frac{d^2y}{dx^2}, \frac{d^3y}{dx^3}, \ldots, \frac{d^ny}{dx^n}, \ldots$

Sample Differentiation: Composite, Implicit, and Inverse Functions Questions

Do not use a calculator for the following two problems.

Suppose $H(x)$ is a differentiable function. Which of the following equals $\frac{d}{dx}\left[\sqrt{x} \cdot H(3x)\right]$?

A. $3\sqrt{x} \cdot H'(3x) + \frac{H(3x)}{2\sqrt{x}}$

B. $\frac{3H'(3x)}{2\sqrt{x}}$

C. $\frac{H(3x)}{2\sqrt{x}} + \sqrt{x} \cdot H'(3x)$

D. $\frac{2}{3}\sqrt[2]{x^3} \cdot H'(3x) + \frac{1}{2\sqrt{x}} \cdot H(3x)$

The correct answer is A. First and foremost, it is important to note that in this scenario, the utilization of the product rule is necessary. When differentiating H(3x), it is important to apply the chain rule:

$$\begin{aligned}\frac{d}{dx}\left[\sqrt{x} \cdot H(3x)\right] &= \sqrt{x} \cdot \left(\frac{d}{dx} H(3x)\right) + \left(\frac{d}{dx}\sqrt{x}\right) \cdot H(3x) \\ &= \sqrt{x} \cdot 3H'(3x) + \frac{1}{2}x^{-1/2} \cdot H(3x) \\ &= 3\sqrt{x} \cdot H'(3x) + \frac{H(3x)}{2\sqrt{x}}\end{aligned}$$

Let m be a nonzero real number and let $H(x) = x^{m^2}$. Which of the following equals $H'''(x)$?

A. $H'''(x) = m^2\left(m^2 - 1\right)\left(m^2 - 2\right)x^{m^2 - 3}$

B. $H'''(x) = \frac{x^{m^2+3}}{\left(m^2+1\right)\left(m^2+2\right)\left(m^2+3\right)}$

C. $H'''(x) = (2m)^3 x^{m^2}$

D. $H'''(x) = m^2\left(m^{2-1}\right)\left(m^{2-2}\right)x^{m^{2-3}}$

The correct answer is A. Apply the power rule three times successively to compute $H'''(x)$:

$$H'(x) = m^2 x^{m^2-1}$$
$$H''(x) = m^2\left(m^2-1\right)x^{m^2-2}$$
$$H'''(x) = m^2\left(m^2-1\right)\left(m^2-2\right)x^{m^2-3}$$

You may use a graphing calculator to solve the following problem.

Which of the following is a complete list of x-values at which the function $f(x) = \ln\left(\left|\sin 2x\right|\right)$ has an irremovable discontinuity?

A. $\varnothing$
B. $x = 2n\pi$, where n is an integer
C. $x = \frac{n\pi}{2}$, where n is an integer
D. $x = n\pi$, where n is an integer

The correct answer is C. The function $y = \ln(u)$ has a vertical asymptote when $u = 0$, and this is an irremovable discontinuity. So, $f(x)$ will have a vertical asymptote at any x-value for which $\sin(2x) = 0$. Observe that $\sin(2x) = 0$ whenever $2x = n\pi$, where n is an integer. This is equivalent to saying $x = \frac{n\pi}{2}$, where n is an integer.

Contextual Applications of Differentiation

About 10–15% of questions on the exam will cover **Contextual Applications of Differentiation.**

In any context, the derivative of a function can be interpreted as the instantaneous rate of change of the independent variable with respect to the dependent variable. If $y = f(x)$, then the units of the derivative are the units of y divided by the units of x.

Straight-Line Motion

Rectilinear (straight-line) motion is described by a function and its derivatives.
If the function $s(t)$ represents the position along a line of a particle at time t, then the velocity is given by $v(t) = s'(t)$. When the velocity is positive, the particle is moving to the right; when it is negative, the particle is moving to the left. The speed of the particle does not take direction into account, so it is the absolute value of the velocity, or $|v(t)|$.

The acceleration of the particle is $a(t) = v'(t) = s''(t)$. The velocity is increasing when $a(t)$ is positive and decreasing when $a(t)$ is negative. The speed, however, is only increasing when $v(t)$ and $a(t)$ have the same sign (positive or negative). When $v(t)$ and $a(t)$ have different signs, the particle's speed is decreasing.

Related Rates

Related rates problems involve multiple quantities that are changing in relation to each other. Derivatives, and especially the chain rule, are used to solve these problems. Though the problems vary widely with context, there are a few steps that usually lead to a solution.

1. Draw a picture and label relevant quantities with variables.
2. Express any rates of change given in the problem as derivatives.
3. Express the rate of change you need to solve for as a derivative.
4. Relate the variables involved in the rates of change to each other with an equation.
5. Differentiate both sides of the equation with respect to time. This may involve applying many derivative rules but will always involve the chain rule.
6. Substitute all of the given information into the resulting equation.
7. Solve for the unknown rate.

Example

The length of the horizontal leg of a right triangle is increasing at a rate of 3 ft/sec, and the length of the vertical leg is decreasing at a rate of 2 ft/sec. At the instant when the horizontal leg is 7 ft and the vertical leg is 1 ft, at what rate is the length of the hypotenuse changing? Is it increasing or decreasing?

We will follow the steps given above.

1.

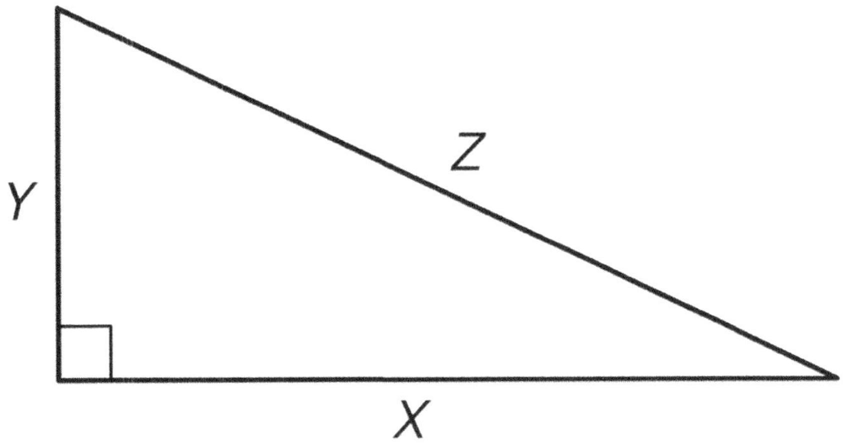

Y

Z

X

2. We are given $\frac{dx}{dt} = 3$ and $\frac{dy}{dt} = -2$.
3. We need to find $\left.\frac{dz}{dt}\right|_{x=7, y=1}$.
4. x, y, and z are related by the Pythagorean theorem: $x^2 + y^2 = z^2$.
5. Differentiating both sides of the equation and applying the chain rule (since all of the variables are functions of t), we get $2x\frac{dx}{dt} + 2y\frac{dy}{dt} = 2z\frac{dz}{dt}$.
6. After substituting all of the information we have, including $x = 7$, $y = 1$, and $z = \sqrt{7^2 + 1^2} = \sqrt{50}$, the equation becomes $2(7)(3) + 2(1)(-2) = 2\left(\sqrt{50}\right)\frac{dz}{dt}$.
7. Solving, we get $\frac{dz}{dt} = \frac{19}{\sqrt{50}}$. The length of the hypotenuse is increasing since its derivative is positive, and it is doing so at a rate of $\frac{19}{\sqrt{50}}$ ft/sec.

Linearization

The line tangent to a function at $x = c$ is the best possible linear approximation to the function near $x = c$. Because of this, the tangent line, seen as a function $L(x)$, is also called the linearization of the function at the given point.

Example

We can use the linearization of $f(x)=3xe^{-x^2}$ at $x = 0$ to approximate the value of $f(0.1)$. To do this, we need to first find the derivative.

Applying the product and chain rules, we get $f'(x)=3\cdot e^{-x^2}+3x\cdot e^{-x^2}\cdot -2x=3e^{-x^2}-6x^2e^{-x^2}$. The slope of the tangent line at $x=0$ is $f'(0)=3e^0-6(0)e^0=3$. The function passes through the point $(0,f(0))=(0,0)$, so the tangent line is $y-0=3(x-0)$.

The linearization of f at $x=0$ is $L(x)=3x$, so the approximation of $f(0.1)$ is $L(0.1)=3(0.1)=0.3$.
Note that the true value of $f(0.1)$ is approximately 0.297, so the linear approximation was an overestimate.

L'Hospital's Rule

When two functions f and g either both have limits of 0 or both have infinite limits, we say that the limit of their ratio is an indeterminate form, represented by $\frac{0}{0}$ or $\frac{\infty}{\infty}$. Limits that result in one of these forms can be evaluated using L'Hospital's rule. The full statement of L'Hospitals rule is as follows: if $\lim_{x\to c}\frac{f(x)}{g(x)}$ approaches $\frac{0}{0}$ or $\frac{\infty}{\infty}$, then $\lim_{x\to c}\frac{f(x)}{g(x)}=\lim_{x\to c}\frac{f'(x)}{g'(x)}$. In other words, when we encounter one of these indeterminate forms, we can take the derivative of each of the functions, and then reevaluate the limit.

Free Response Tip

Free response questions frequently feature limits that necessitate the use of L'Hospital's Rule. It's important to avoid mixing up L'Hospital's Rule and the quotient rule. The derivative of the ratio is not being computed by taking the derivative of the numerator and denominator separately.

Practice Contextual Applications of Differentiation Questions

Do not use a calculator for the following two problems.

Compute the limit $\lim\limits_{x \to 0^+} \dfrac{\ln(\tan x)}{\ln(\sin x)}$.

A. –1
B. 0
C. 1
D. ∞

The correct answer is C. Plugging $x = 0$ into the expression shows that the limit is indeterminate of the form $\frac{0}{0}$. So, use l'Hopital's rule:

$$\lim_{x \to 0^+} \frac{\ln(\tan x)}{\ln(\sin x)} = \lim_{x \to 0^+} \frac{\frac{1}{\tan x} \cdot \sec^2 x}{\frac{1}{\sin x} \cdot \cos x} = \lim_{x \to 0^+} \frac{\cancel{\cot x} \cdot \sec^2 x}{\cancel{\cot x}} = \lim_{x \to 0^+} \sec^2 x = \sec^2(0) = 1$$

To prevent the balloon from bursting, it is necessary to inflate the spherical Mylar balloon at a rate of 2 cubic inches per second, ensuring a steady increase in its volume. What is the rate of increase in the diameter when it reaches a size of 4 inches?

A. $\frac{3}{16\pi}$ inches per second
B. $\frac{1}{32\pi}$ inches per second
C. $\frac{1}{4\pi}$ inches per second
D. $\frac{1}{8\pi}$ inches per second

The correct answer is C. Differentiate the volume formula $V = \frac{4}{3}\pi r^3$ with respect to t:

$$\frac{dV}{dt} = \frac{4}{3} \cdot 3\pi r^2 \cdot \frac{dr}{dt} = 4\pi r^2 \cdot \frac{dr}{dt}$$

Now, substitute in the known information:

$$2 \ {}^{\text{inches}^3}\!/_{\text{sec}} = 4\pi(2 \text{ inches})^2 \cdot \frac{dr}{dt}$$

$$\frac{dr}{dt} = \frac{1}{8\pi} \ {}^{\text{inches}}\!/_{\text{sec}}$$

Since $D = 2r$, it follows that $\frac{dD}{dt} = 2 \cdot \frac{dr}{dt} = \frac{1}{4\pi}$ $^{\text{inches}}/_{\text{sec}}$.

You may use a graphing calculator to solve the following problem.

An object moves along a number line and its position at time t is given by $p(t) = t\cos(3t),\ t \geq 0$. What is the first-time interval, approximately, on which the speed of the object is increasing?

A. (0, 0.293)
B. (0.623, 0.755)
C. (0.291, 1.162)
D. (0.755, 1.713)

The correct answer is B. The speed function is $|p'(t)|$, which is given by

$$|p'(t)| = |t(-3\sin(3t)) + \cos(3t)| = |-3t\sin(3t) + \cos(3t)|$$

Use the graphing calculator to get the following graph:

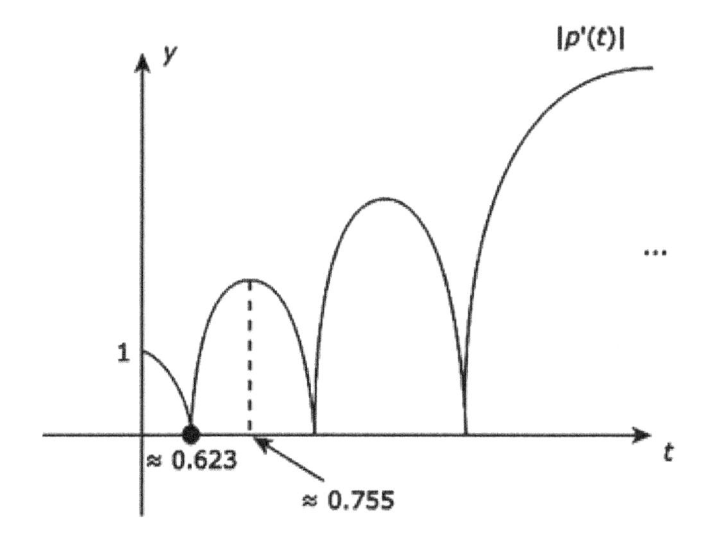

Observe that the first interval on which this graph is increasing is (0.623, 0.755).

Analytical Applications of Differentiation

A significant portion of the questions on your AP exam will focus on Analytical Applications of Differentiation, ranging from 15–18%.

Mean Value Theorem

The Mean Value Theorem states that if f is continuous on $[a,b]$ and differentiable on (a,b), then there is at least one point between a and b at which the instantaneous rate of change of f is equal to its average range of change over the entire interval. In other words, there is at least one value c in the interval (a,b) for which $f'(c)=\dfrac{f(b)-f(a)}{b-a}$.

Example

Let $f(x)=x^2$. Over the interval [3,7], the average rate of change of f is $\dfrac{f(7)-f(3)}{7-3}=\dfrac{40}{4}=10$. Since f is continuous and differentiable everywhere, the Mean Value Theorem guarantees that there is at least one c between 3 and 7 for which $f'(c)=10$. Since $f'(x)=2x$, we can find the guaranteed value(s) of c by solving $2x = 10$. In this case, of course, there is exactly one such value: $x = 5$.

Free Response Tip

When encountering the phrase "explain why there must be a value..." in a free response question, it is important to consider two theorems. When dealing with a function that represents a derivative, it's important to start by considering the Mean Value Theorem. If not, you might want to consider the Intermediate Value Theorem. Regardless of the situation, it is essential to provide a solid rationale for the application of thc theorem based on the principles of continuity and differentiability.

Intervals of Increase and Decrease and the First Derivative Test

When the derivative of a function is positive, the function experiences an increase, and when the derivative is negative, the function undergoes a decrease. In order to determine the intervals on which a function is increasing or decreasing, it is essential to solve for the points where its derivative is positive or negative. The process for accomplishing this requires initially identifying the values, known as critical points, where the derivative is zero or undefined.

If f changes from increasing to decreasing at $x = c$, f has a local maximum at c. If it changes from decreasing to increasing at $x = c$, it has a local minimum at c. Taken together, local maximums and local minimums are referred to as local extrema.

The first derivative test summarizes these facts and describes the process of finding local maximums and minimums. Specifically, suppose $x = c$ is a critical point of f. Then:

- if f' is positive to the left of c, and negative to the right of c, then f has a local maximum at c.
- if f' is negative to the left of c, and positive to the right of c, then f has a local minimum at c.
- if neither of the above conditions apply, f does not have a local extreme at c.

Example

Let $f(x) = x^5 - 3x^3$. To find the local extrema of f, we begin by finding the derivative, setting it to 0, and solving for x:

$$f'(x) = 5x^4 - 9x^2$$

$$5x^4 - 9x^2 = 0$$

$$x^2\left(5x^2 - 9\right) = 0$$

$$x = 0, \frac{3}{\sqrt{5}}, -\frac{3}{\sqrt{5}}$$

Since f' is never undefined, these three values are the only critical points of f. These critical points divide the real number line into four intervals: $\left(-\infty, -\frac{3}{\sqrt{5}}\right)$, $\left(-\frac{3}{\sqrt{5}}, 0\right)$, $\left(0, \frac{3}{\sqrt{5}}\right)$, and $\left(\frac{3}{\sqrt{5}}, \infty\right)$. From each of these intervals we choose a point and use it to determine whether f' is positive or negative on the interval. Note that $\frac{3}{\sqrt{5}} \approx 1.34$.

	$\left(-\infty, -\frac{3}{\sqrt{5}}\right)$	$\left(-\frac{3}{\sqrt{5}}, 0\right)$	$\left(0, \frac{3}{\sqrt{5}}\right)$	$\left(\frac{3}{\sqrt{5}}, \infty\right)$
Test point $x = a$	-2	-1	1	2
$f'(a)$	$f'(-2) = 44$	$f'(-1) = -4$	$f'(1) = -4$	$f'(2) = 44$
Conclusion	f' is positive, so f is increasing	f' is negative, so f is decreasing	f' is negative, so f is decreasing	f' is positive, so f is increasing

Examining the table above, we see that f changes from increasing to decreasing at $x = -\frac{3}{\sqrt{5}}$, so f has a local maximum there. Also, f changes from decreasing to increasing at $x = \frac{3}{\sqrt{5}}$, so f has a local minimum there. Note that at $x = 0$, f has neither a maximum nor a minimum, since the derivative does not change sign from the left to the right of the point.

Absolute Extrema

If $f(c) = M$ is the largest value that f attains on some interval I containing c, then M is called the global maximum of f on I. Similarly, if $f(c) = M$ is the smallest value that f attains on some interval I containing c, then M is called the global minimum of f on I.

It is not guaranteed that any random function will have a global maximum or minimum value on a given interval. Nevertheless, the Extreme Value Theorem ensures that a function will always possess a global maximum and a global minimum on any closed interval where it remains continuous. During this interval, the global extrema will occur either at a critical point or at one of the endpoints.

The Candidate Test gives a procedure for finding these global extrema on a closed interval $[a,b]$:

1. Check that f is continuous on $[a,b]$.
2. Find the critical numbers of f between a and b.
3. Check the value of f at each critical number, at a, and at b.
4. The largest value found in the previous step is the global maximum, and the smallest value found is the global minimum.

Example

Let us find the global extrema of $f(x) = 2x^3 + 3x^2 - 12x - 1$ on [–1,3] by following the steps given.

First, note that *f* is a polynomial function, so it is continuous everywhere. The derivative of *f* is $f'(x) = 6x^2 + 6x - 12$. This is always defined, so we need only set it to 0 and solve:

$$6x^2 + 6x - 12 = 0$$
$$6(x+2)(x-1) = 0$$
$$x = -2, 1$$

$x = -2$ is not in [–1, 3], so we will only consider $x = 1$. Now we will check the value of *f* at this critical point and at the endpoints of the interval.

-1	$f(-1) = 2(-1)^3 + 3(-1)^2 - 12(-1) - 1 = 12$
1	$f(1) = 2(1)^3 + 3(1)^2 - 12(1) - 1 = -8$
3	$f(3) = 2(3)^3 + 3(3)^2 - 12(3) - 1 = 44$

The maximum value of *f* on [–1, 3] is 44, and it occurs at the endpoint $x = 3$. The minimum value is –8, and it occurs at the critical point $x = 1$.

Concavity and Inflection Points

The graph of a function *f* is concave up when its derivative f' is increasing, and it is concave down when f' is decreasing. Since the relationship of f'' to f' is the same as the relationship of f' to *f*, we can determine on which intervals f' is increasing (or decreasing) by checking where f'' is positive (or negative). Therefore, the criteria for *f* being concave up or down can be restated in terms of f'': *f* is concave up when f'' is positive, and concave down when f'' is negative.

A point where a function changes concavity is known as a point of inflection. These can be discovered in a similar manner to how local extrema are located using the first derivative test: identify where the second derivative is 0 or undefined, and examine points on either side to determine if concavity is changing.

Example

The function in the previous example, $f(x) = 2x^3 + 3x^2 - 12x - 1$, has second derivative $f''(x) = 12x + 6$. This is defined everywhere and is 0 only at $x = -\frac{1}{2}$. To the left of $-\frac{1}{2}$, say at $x = -1$, we have $f''(-1) = 12(-1) + 6 = -6$, which is negative. Therefore, *f* is concave down to the left of $-\frac{1}{2}$. To the right of $-\frac{1}{2}$, say at $x = 0$, we have $f''(0) = 12(0) + 6 = 6$, which is positive. This means that *f* is concave up to the right of $-\frac{1}{2}$. Since *f* changes from concave down to concave up as it passes through $-\frac{1}{2}$, *f* has a point of inflection at $x = -\frac{1}{2}$.

Second Derivative Test

In addition to providing information about concavity and inflection points, the second derivative of a function can also help determine whether a critical point represents a relative maximum or minimum. Specifically, suppose *f* has a critical point at $x = c$. Then:

- if $f''(c) > 0$, *f* has a local minimum at *c*.
- if $f''(c) < 0$, *f* has a local maximum at *c*.
- if $f''(c) = 0$, this test is inconclusive, and the first derivative test must be used.

Summary of Curve Sketching

The following table summarizes the behavior of a graph at $x = c$, depending on the values of $f'(c)$ and $f''(c)$.

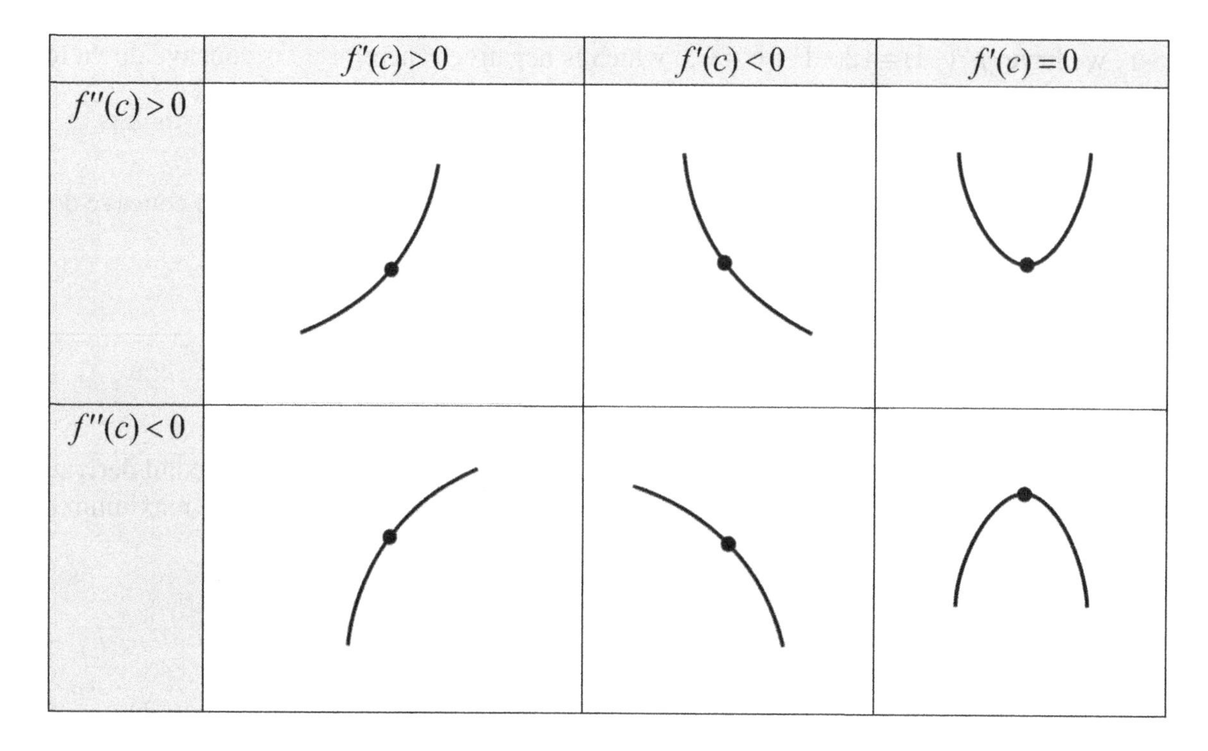

Optimization

The methods provided for determining local and global extrema can be utilized in a diverse range of application problems, commonly referred to as optimization problems. The specific steps and approach may differ depending on the context, but there are generally accepted guidelines for handling these types of situations:

1. Draw a picture.
2. Write a function for the quantity to be optimized (maximized or minimized).
3. Rewrite the function from the previous step to be in terms of a single independent variable. This often involves using a secondary equation, called a constraint.
4. Determine the domain of interest.
5. Differentiate the function and find the relevant critical points.
6. Use the first derivative test, second derivative test, or candidates test to determine which of the critical points or endpoints represent the optimal solution.

Example

A manufacturer wants to construct a cylindrical container with a volume of 5 ft^3. Using the steps noted previously, let's find the dimensions of the container that will minimize the amount of material used.

1.

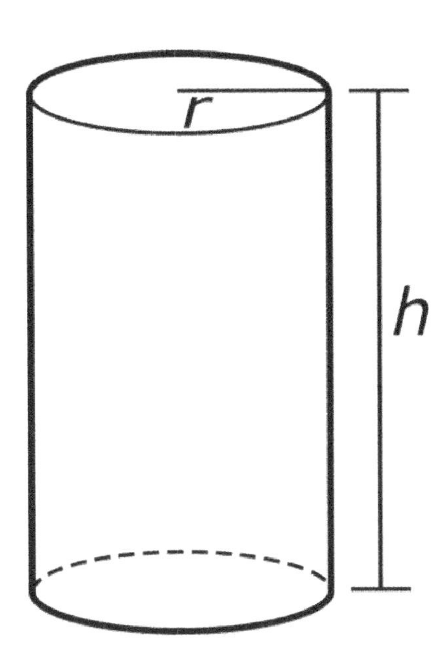

2. The quantity to be optimized is the surface area of the container. In terms of r and h, the surface area is given by the function $S = 2\pi r^2 + 2\pi rh$.
3. As written, the function that gives the surface area depends on both r and h. However, since we know the volume of the cylinder is to be 5, and the volume formula is $V = \pi r^2 h$, we have the constraint $\pi r^2 h = 5$. Solving for h gives $h = \dfrac{5}{\pi r^2}$. This can be substituted into the function S:

   $S = 2\pi r^2 + 2\pi rh$

   $S = 2\pi r^2 + 2\pi r\left(\dfrac{5}{\pi r^2}\right)$

   $S = 2\pi r^2 + \dfrac{10}{r}$

   Now S is written in terms of a single variable, r.
4. Considering the physical situation, it is clear that the domain of interest is $r > 0$. A cylinder cannot exist with $r \leq 0$.
5. Differentiating and setting to zero:

$$\frac{dS}{dr} = 4\pi r - \frac{10}{r^2}$$
$$4\pi r - \frac{10}{r^2} = 0$$
$$4\pi r^3 - 10 = 0$$
$$r^3 = \frac{5}{2\pi}$$
$$r = \sqrt[3]{\frac{5}{2\pi}}$$
The derivative is undefined at $r = 0$, but that is irrelevant since it is not in the domain.

6. The only critical point is $r = \sqrt[3]{\frac{5}{2\pi}}$, and the domain of r is $(0, \infty)$, so there are no endpoints. To justify that this critical point is indeed a minimum, we will use the second derivative test.
   $\frac{d^2S}{dr^2} = 4\pi + \frac{20}{r^3}$. Evaluating at $r = \sqrt[3]{\frac{5}{2\pi}}$, we have $\frac{d^2S}{dr^2} = 4\pi + \frac{20}{\left(\sqrt[3]{\frac{5}{2\pi}}\right)^3} = 12\pi$. Since this is positive, the critical point is indeed a minimum, as desired.

> **Free Response Tip**
>
> Similar to the previous example, numerous applied optimization problems seem to have a single viable solution. Regardless of the situation, it is important to provide a rationale for why this solution is considered the optimal choice. One approach to solving this problem is to consider the Second Derivative Test, which is often the most straightforward method. However, it's important to also remember the First Derivative Test and the Candidates Test as alternative options.

Implicitly Defined Curves

When a curve is defined implicitly in an equation involving x and y, the applications of derivatives discussed in this section still generally apply. As with explicitly defined functions, critical points are determined by examining where $\frac{dy}{dx} = 0$ or is undefined. However, the details of finding where this occurs are often more complicated since the expression for $\frac{dy}{dx}$ usually involves both x and y. Second derivatives are often trickier to find as well. Two points are helpful:

- The derivative of $\frac{dy}{dx}$ with respect to x is the second derivative of y with respect to x. In other words, $\frac{d}{dx}\left(\frac{dy}{dx}\right)=\frac{d^2y}{dx^2}$.
- When the expression for $\frac{d^2y}{dx^2}$ involves $\frac{dy}{dx}$, it is usually possible to simplify by substituting a previously obtained expression for $\frac{dy}{dx}$.

Example

Suppose $x^2+2xy=0$. The derivative $\frac{dy}{dx}$ can be found by differentiating with respect to x and solving for $\frac{dy}{dx}$:

$$\frac{d}{dx}\left(x^2+2xy\right)=\frac{d}{dx}(0)$$

$$2x+2\cdot y+2x\cdot 1\frac{dy}{dx}=0$$

$$\frac{dy}{dx}=\frac{-x-y}{x}$$

To find the second derivative, differentiate both sides of this result with respect to x:

$$\frac{d}{dx}\left(\frac{dy}{dx}\right)=\frac{d}{dx}\left(\frac{-x-y}{x}\right)$$

$$\frac{d^2y}{dx^2}=\frac{\left(-1-1\frac{dy}{dx}\right)(x)-(-x-y)(1)}{x^2}$$

Now substituting $\frac{-x-y}{x}$ for $\frac{dy}{dx}$:

$$\frac{d^2y}{dx^2}=\frac{\left(-1-1\left(\frac{-x-y}{x}\right)\right)(x)-(-x-y)(1)}{x^2}$$

$$\frac{d^2y}{dx^2}=\frac{-x+x+y+x+y}{x^2}$$

$$\frac{d^2y}{dx^2}=\frac{x+2y}{x^2}$$

Sample Analytical Applications of Differentiation Questions

Do not use a calculator for the following two problems.

Consider the function $g(x) = -x^2 + 2x + 1$ on the interval [–1,2]. Which of the following values of c, if any, satisfies the conclusion of the Mean Value Theorem on the interval [–1, 2]?

A. 0
B. $\frac{1}{2}$
C. 1
D. No such c value exists

The correct answer is B. Observe that $g(x)$ is continuous on [–2,1] and differentiable on (–2,1). So, the Mean Value Theorem guarantees the existence of at least one value of c in (–2,1) for which $g'(c) = \frac{g(2) - g(-1)}{2 - (-1)}$. Since for this function $g'(x) = -2x + 2$, this condition reduces to

$$-2c + 2 = \frac{1 - (-2)}{2 - (-1)} = 1.$$

Solving for c yields $c = \frac{1}{2}$.

So, you're looking to build a rectangular box with a square base and a volume of 3200 cubic inches. The material for the top and bottom is priced at $1.50 per square inch, while the material for the sides is priced at $2.75 per square inch. Which of the functions below would you optimize to determine the dimensions that would yield the least cost?

A. $C(x) = 3x^2 + \frac{35,200}{x}$

B. $C(x) = 2x^2 + \frac{12,800}{x}$

C. $C(x) = \frac{3}{2}x^2 + \frac{3200}{x}$

D. $C(x) = 1.5x^2 + \frac{8800}{x}$

The correct answer is A. Let x = width of the base = length of the base, and y = height of the box.

The volume is given by $x^2 y = 3200$, so $y = \frac{3200}{x^2}$. The cost of construction is linked to the surface area formula for the box:

Face of Box	Area	Cost
Top	x^2	$\$1.50x^2$
Bottom	x^2	$\$1.50x^2$
Each Lateral Face	xy	$\$2.75xy$

So, the total cost of the construction is $2(1.50x^2) + 4(2.75xy)$. Substituting $y = \frac{3200}{x^2}$ to give the following cost function in x that should be minimized: $C(x) = 3x^2 + 11x\left(\frac{3200}{x^2}\right) = 3x^2 + \frac{35,200}{x}$.

You may use a graphing calculator to solve the following problem.

The derivative of a function $f(x)$ is given by $f'(x) = \frac{1}{3} + \cos\left(x^3\right) - 2\ln x$. At approximately what x-value in the interval $(0, \infty)$ does $f(x)$ have a local maximum value?

A. 1.181
B. 1.500
C. 1.783
D. 1.881

The correct answer is C. Use the graphing calculator to graph $f'(x) = \frac{1}{3} + \cos\left(x^3\right) - 2\ln x$ on a small interval, say [0,2]:

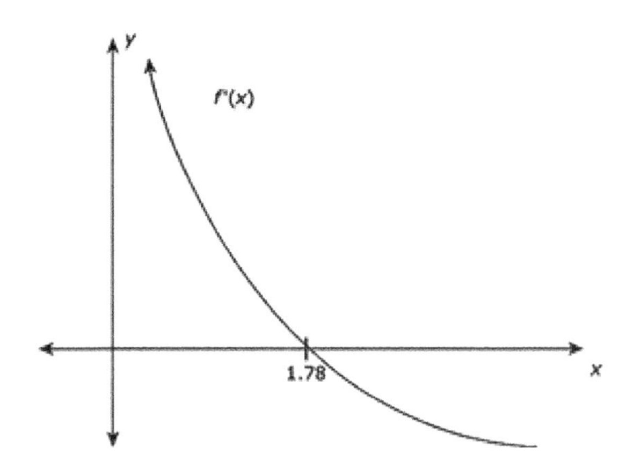

So, 1.783 is the approximate x-value at which $f(x)$ has a local maximum because the graph of $f'(x)$ is positive to its left close by and positive to its right close by.

Integration and Accumulation of Change

Approximately 17–20% of the questions on your AP Calculus AB exam will focus on Integration and Accumulation of Change.

Riemann Sums and the Definite Integral

When a function represents a rate of change, the region bounded by the graph of the function and the x-axis signifies the accumulation of the change. When the area is located above the x-axis, the accumulated change becomes positive. Conversely, if the area is below the x-axis, the accumulated change turns negative.

More generally, the accumulation of a function on a closed interval $[a, b]$, represented graphically by the area between a function and the x-axis, is called the definite integral of the function on that interval, and is denoted $\int_a^b f(x)\,dx$.

For simple functions, the definite integral can often be evaluated geometrically.

Example

To evaluate $\int_{-1}^{4}(x-3)\,dx$, draw a picture:

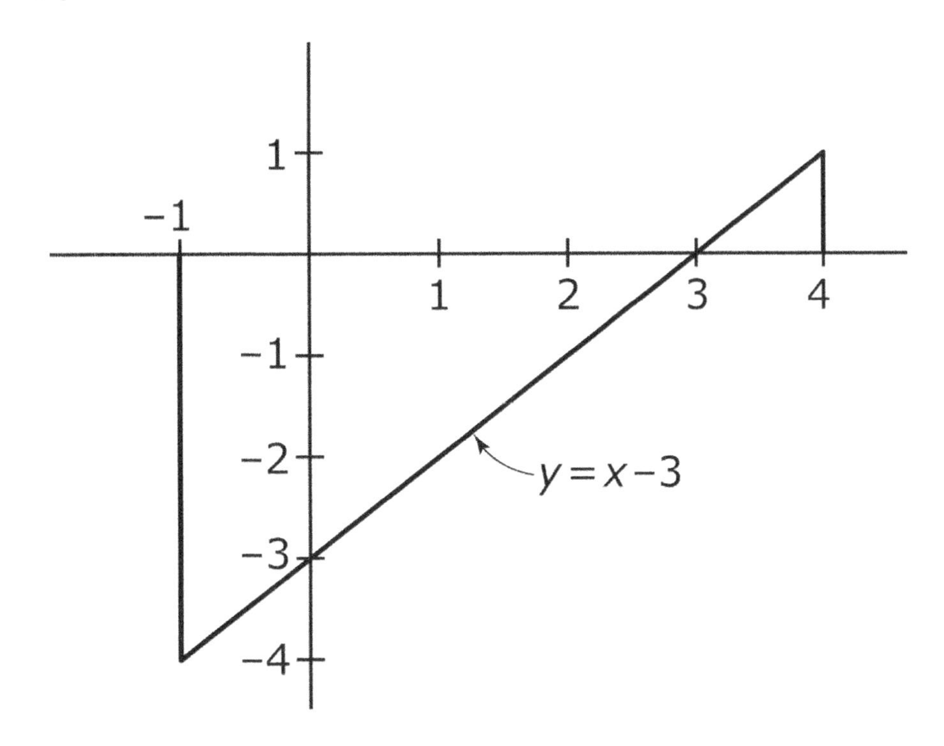

The area between the curve and the graph is divided into two triangles. The larger triangle has an area of 8. However, it is below the x-axis, so the accumulation is of negative values. Therefore, it contributes a value of -8 to the integral. The smaller triangle accumulates positive values and has an area of $\frac{1}{2}$. Together, we have $\int_{-1}^{4}(x-3)\,dx=-8+\frac{1}{2}=-\frac{15}{2}$.

Definite integrals can be approximated using a variety of sums, each term of which represents the area of a rectangle over a small subinterval. To begin, consider a function $f(x)$ over the interval $[a,b]$, and let n be the number of equally sized subintervals into which it is split. Then $\Delta x=\frac{b-a}{n}$ is the width of each subinterval, and $x_i=a+i\Delta x$ is the left endpoint of the ith subinterval. If a rectangle is constructed on each subinterval so that its height is equal to the value of $f(x_i)$, the sum of the areas will be $\left(f(x_0)+f(x_1)+\cdots+f(x_{n-1})\right)\Delta x$. This sum is called a left Riemann sum.

The notation $\sum_{i=1}^{n}a_i$ stands for the sum $a_1+a_2+\cdots+a_n$. The left Riemann sum can be written using this notation as $\sum_{i=0}^{n-1}f(x_i)\Delta x$. Other commonly used approximations are the right Riemann sum, and the midpoint Riemann sum, as shown below.

Left Riemann Sum	Right Riemann Sum	Midpoint Riemann Sum
$\sum_{i=0}^{n-1}f(x_i)\Delta x$	$\sum_{i=1}^{n}f(x_i)\Delta x$	$\sum_{i=0}^{n-1}f\left(\frac{x_i+x_{i+1}}{2}\right)\Delta x$

As n increases in size, each of these Riemann sums becomes a more accurate approximation of the definite integral. When the limit is taken as $n\to\infty$, any of these sums becomes equal to the definite integral. In other words, the integral of a function f over the integral $[a,b]$ can be defined as $\int_a^b f(x)\,dx=\lim_{n\to\infty}\sum_{i=0}^{n-1}f(x_i)\Delta x$, provided this limit exists.

In fact, although $\Delta x=\frac{b-a}{n}$ is the most common way to divide intervals into subintervals, all of the above sums can be computed with potentially different Δx values for each subinterval. The limit of the sum is still equal to the definite integral.

Another expression that can be used to approximate the definite integral is a trapezoidal sum, which represents the areas of trapezoids, rather than rectangles, constructed over the subintervals. The trapezoidal sum is $\frac{\Delta x}{2}\left(f(x_0)+2f(x_1)+2f(x_2)+\cdots+2f(x_{n-1})+f(x_n)\right)$.

Free Response Tip

Values of a function in a table are commonly found in free response questions. An approximation of the integral can be obtained by utilizing a Riemann sum, which takes into account the subintervals provided in the table, even if their lengths vary. In the process of calculating Riemann sums, it is important to consider the length of each subinterval, which represents the distance between consecutive x-values. Additionally, the height of the rectangle on each subinterval is determined by the y-value associated with either the left x (for a left Riemann sum) or the right x (for a right Riemann sum). Regardless of the situation, the sum you have should have one less term than the number of points provided in the table.

Properties of the Definite Integral

The definite integral satisfies several properties:

- $\int_a^b c\,dx = c(b-a)$, for any constant c
- $\int_a^b cf(x)\,dx = c\int_a^b f(x)\,dx$
- $\int_a^b [f(x) \pm g(x)]\,dx = \int_a^b f(x)\,dx \pm \int_a^b g(x)\,dx$
- $\int_a^a f(x)\,dx = 0$
- $\int_a^b f(x)\,dx = -\int_b^a f(x)\,dx$
- $\int_a^b f(x)\,dx = \int_a^c f(x)\,dx + \int_c^b f(x)\,dx$

Example

Suppose $\int_1^7 f(x)\,dx = 9$ and $\int_1^4 f(x)\,dx = 12$. Find $\int_7^4 (f(x)-3)\,dx$.

First, we have $\int_7^4 (f(x)-3)\,dx = \int_7^4 f(x)\,dx - \int_7^4 3\,dx$. The latter integral is simply $3(4-7) = -9$.

For the former:

$$\begin{aligned}\int_7^4 f(x)\,dx &= \int_7^1 f(x)\,dx + \int_1^4 f(x)\,dx \\ &= -\int_1^7 f(x)\,dx + 12 \\ &= -9 + 12 \\ &= 3\end{aligned}$$

The answer is then $\int_7^4 (f(x)-3)\,dx = 3-(-9) = 12$.

Accumulation Functions and the Fundamental Theorem of Calculus

A function can be defined in terms of a definite integral: $g(x) = \int_a^x f(t)\,dt$. The first part of the Fundamental Theorem of Calculus states that the derivative of this function at a given point is equal to the value of the function being accumulated. That is, $g'(x) = \frac{d}{dx}\left(\int_a^x f(t)\,dt\right) = f(x)$.

Since $\int_x^a f(t)\,dt = -\int_a^x f(t)\,dt$, we also have $\frac{d}{dx}\left(\int_x^a f(t)\,dt\right) = -\frac{d}{dx}\left(\int_a^x f(t)\,dt\right) = -f(x)$. If the upper limit of integration is a function of *x*, the chain rule can be applied along with the fundamental theorem.

Example

If $f(x) = \int_2^{x^2} \sin t\,dt$, then $f'(x) = \sin(x^2)\cdot 2x$.

Antiderivatives and the Fundamental Theorem of Calculus

If $g'(x) = f(x)$, *g* is said to be an antiderivative of *f*. Note that if *C* is any constant, then $\frac{d}{dx}[g(x)+C] = g'(x)+0 = f(x)$, so that $g(x)+C$ is also an antiderivative of *f*. In fact, all antiderivatives of a given function have this relationship with each other: they differ only by a constant. Every continuous function *f* has an antiderivative, since the function $g(x) = \int_a^x f(t)\,dt$ satisfies $g'(x) = f(x)$ and is therefore an antiderivative of *f*.

The second part of the Fundamental Theorem of Calculus states that if *f* is continuous on the interval $[a,b]$, and *F* is any antiderivative of *f* on that interval, then $\int_a^b f(x)\,dx = F(b) - F(a)$.

This fact means that antiderivatives and integrals are very closely related. Because of this, an antiderivative is also called an indefinite integral, and is denoted $\int f(x)\,dx = F(x) + C$, where F is any antiderivative.

Basic Rules of Antiderivatives

Since finding an antiderivative is the inverse process of finding a derivative, the rules for derivatives can be reversed to find antiderivatives.

$f(x)$	□$f(x)\,dx$
x^n	$\frac{1}{n+1}x^{n+1} + C$
e^x	$e^x + C$
$\frac{1}{x}$	$\ln\lvert x\rvert + C$
$\sin x$	$-\cos x + C$
$\cos x$	$\sin x + C$
$\sec^2 x$	$\tan x + C$
$\sec x \tan x$	$\sec x + C$
$\csc x \cot x$	$-\csc x + C$
$\csc^2 x$	$-\cot x + C$

Integration by Substitution

Substitution, also known as change of variables, is a technique for finding antiderivatives and is analogous to the chain rule for derivatives. It works by noting that $\int f'(g(x))g'(x)\,dx = f(g(x)) + C$. The technique, then, requires recognizing the $g(x)$ and $g'(x)$ in the expression being integrated.

If $u = g(x)$, then $du = g'(x)\,dx$, so the integral can be written $\int f(u)\,du = f(u) + C$.

Example

Consider $\int x^2 e^{x^3}\,dx$. If we let $u = x^3$, then $du = 3x^2\,dx$, or $x^2\,dx = \frac{1}{3}du$. Since constants can be pulled out of integrals, the integral then becomes $\frac{1}{3}\int e^u\,du = \frac{1}{3}e^u + C = \frac{1}{3}e^{x^3} + C$.

When using this technique with definite integrals, it is important to translate the limits of integration to be in terms of the new function u.

Example

The integral $\int_{\pi/2}^{\pi} \sin^2\theta \cos\theta \, d\theta$ can be evaluated by substituting $u = \sin\theta$. Then $du = \cos\theta \, d\theta$.

When $\theta = \frac{\pi}{2}$, $u = \sin\frac{\pi}{2} = 1$, and when $\theta = \pi$, $u = \sin\pi = -1$. The integral becomes

$\int_1^0 u^2 \, du = \left[\frac{1}{3}u^3\right]_1^0 = 0 - \frac{1}{3} = -\frac{1}{3}$.

Other Integration Techniques

When the top part of a rational function has a degree that is equal to or greater than the bottom part, long division can be quite useful when it comes to integration.

Example

Consider $\int \frac{x^3 + x}{x - 1} dx$. Since the numerator has a higher degree than the denominator, long division can be applied to transform the integral. We get $\frac{x^3 + x}{x - 1} = x^2 + x + 2 + \frac{2}{x - 1}$, so that the answer is $\frac{1}{3}x^3 + \frac{1}{2}x^2 + 2x + 2\ln|x - 1| + C$.

Another technique that can be useful for some integrals is completing the square.

Example

Given $\int \frac{1}{x^2 - 6x + 10} dx$, note that completing the square in the denominator results in $\int \frac{1}{(x-3)^2 + 1} dx$. This should now be recognizable as the derivative of an arctan function, and the antiderivative is $\arctan(x - 3) + C$.

Practice Integration and Accumulation of Change Questions

Do not use a calculator for the following two problems.

Let $y(\theta) = \theta \cdot \int_4^\theta \ln(3t)dt$. Compute $y''(\theta)$.

A. 1
B. $\frac{1}{3} + 2\ln(3\theta)$
C. $\frac{1}{3} + \ln(3\theta)$
D. $1 + 2\ln(3\theta)$

The correct answer is D. To compute $y'(\theta)$, you must first and foremost use the product rule, and then when you differentiate the integral term, use the Fundamental Theorem of Calculus:

$$\begin{aligned} y'(\theta) &= \theta \cdot \tfrac{d}{d\theta}\left(\int_4^\theta \ln(3t)dt\right) + \left(\int_4^\theta \ln(3t)dt\right) \cdot \tfrac{d}{d\theta}\theta \\ &= \theta \cdot \ln(3\theta) + 1 \cdot \left(\int_4^\theta \ln(3t)dt\right) \end{aligned}$$

Next, compute $y''(\theta)$ in a similar fashion:

$$\begin{aligned} y''(\theta) &= \theta \cdot \tfrac{d}{d\theta}\ln(3\theta) + \left(\ln(3\theta)\right) \cdot \tfrac{d}{d\theta}\theta + \tfrac{d}{d\theta}\left(\int_4^\theta \ln(3t)dt\right) \\ &= \theta \cdot \tfrac{1}{3\theta} \cdot 3 + \ln(3\theta) \cdot 1 + \ln(3\theta) \\ &= 1 + 2\ln(3\theta) \end{aligned}$$

Compute $\int_2^x \frac{\ln\left(\frac{t}{2}\right)}{t}\,dt$.

A. $\frac{1}{2}\left(\ln\left(\frac{x}{2}\right)\right)^2$

B. $\left(\ln\left(\frac{x}{2}\right)\right)^2$

C. $\frac{1}{2}\left(\ln\left(\frac{x}{2}\right)\right)^2 - \frac{1}{2}e^2$

D. $\frac{1}{2}x^2 - 2$

The correct answer is A. Use the following u-substitution:

$$u = \ln\left(\tfrac{t}{2}\right)$$

$$du = \frac{1}{\frac{t}{2}} \cdot \tfrac{1}{2}\,dt = \tfrac{1}{t}\,dt$$

$$t = 2 \Rightarrow u = \ln 1 = 0$$

$$t = x \Rightarrow u = \ln\left(\tfrac{x}{2}\right)$$

So, the integral is now evaluated as follows:

$$\int_2^x \frac{\ln\left(\frac{t}{2}\right)}{t}\,dt = \int_0^{\ln\left(\frac{x}{2}\right)} u\,du = \tfrac{1}{2}u^2\Big|_0^{\ln\left(\frac{x}{2}\right)} = \tfrac{1}{2}\left(\ln\left(\tfrac{x}{2}\right)\right)^2$$

You may use a graphing calculator to solve the following problem.

Consider the piecewise–defined function

$$f(x) = \begin{cases} x^3 + 9x^2 + 27x + 19, & x \le -3 \\ \ln\left(|x|\right), & -3 < x < 0 \end{cases}$$

Which of the following expressions gives the area of the region bounded by the graph of $f(x)$, the x-axis, $x = -4$, and $x = -\frac{1}{2}$?

A. $\int_{-3}^{-1} f(x)dx - \left[\int_{-4}^{-3} f(x)dx + \int_{-1}^{-1/2} f(x)dx\right]$

B. $\int_{-3}^{-1/2} f(x)dx - \int_{-4}^{-3} f(x)dx$

C. $\int_{-4}^{-1/2} f(x)dx$

D. $\int_{-1/2}^{-4} f(x)dx$

The correct answer is A. First, graph the region using the graphing calculator:

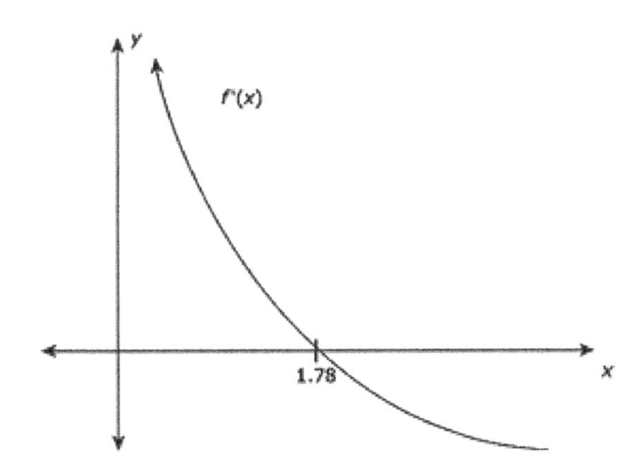

Use interval additivity to break the region into three disjoint pieces on the subintervals [–4,–3], [–3,–1], and [–1, $-\frac{1}{2}$]. Since the regions on [–4, –3] and [–1, $-\frac{1}{2}$] are below the x-axis, we integrate $-f(x)$ on them, while we integrate $f(x)$ on [–3,–1] since the graph is above the x-axis there. Summing these integrals yields the area of the region:

$$\int_{-4}^{-3} -f(x)dx + \int_{-3}^{-1} f(x)dx + \int_{-1}^{-1/2} -f(x)dx = \int_{-3}^{-1} f(x)dx - \left[\int_{-4}^{-3} f(x)dx + \int_{-1}^{-1/2} f(x)dx\right],$$

where linearity was used to get the right-side of the inequality.

Differential Equations

A small portion, approximately 6–12%, of the questions on your exam will focus on Differential Equations.

Introduction to Differential Equations

Differential equations are equations that involve functions and their derivatives. A differential equation is solved by finding a function that meets the requirements of the equation.

Example

Consider the differential equation $y' - 3y - 3 = 0$. One solution to this equation is given by $y = e^{3x} - 1$. To check this, first find $y' = 3e^{3x}$, and now substitute y' and y into the differential equation: $y' - 3y - 1 = 3e^{3x} - 3(e^{3x} - 1) - 3 = 0$.

A differential equation may have infinitely many solutions parameterized by a constant; this is called the general solution to the equation. If additional information is given, the constant can be determined. This additional information comes in the form of an initial condition; that is, a value $f(x_0) = y_0$ that must be satisfied by the solution.

Example

Consider the differential equation $y'' = -y$ with initial condition $y\left(\frac{\pi}{2}\right) = -7$. Any function of the form $y = C\sin x$ is a general solution to this equation since $\frac{d^2}{dx^2}(C\sin x) = -C\sin x$. Using the initial condition given, we have $-7 = C\sin\frac{\pi}{2}$, so we can solve to find $C = -7$. The particular solution to the equation is $y = -7\sin x$.

Slope Fields

A slope field provides a visual depiction of a differential equation. A short line is drawn at each of a limited number of points in a particular section of a plane, representing the slope of a function. This equation represents a differential equation, and its solution is the function that corresponds to the slopes being depicted.

Example

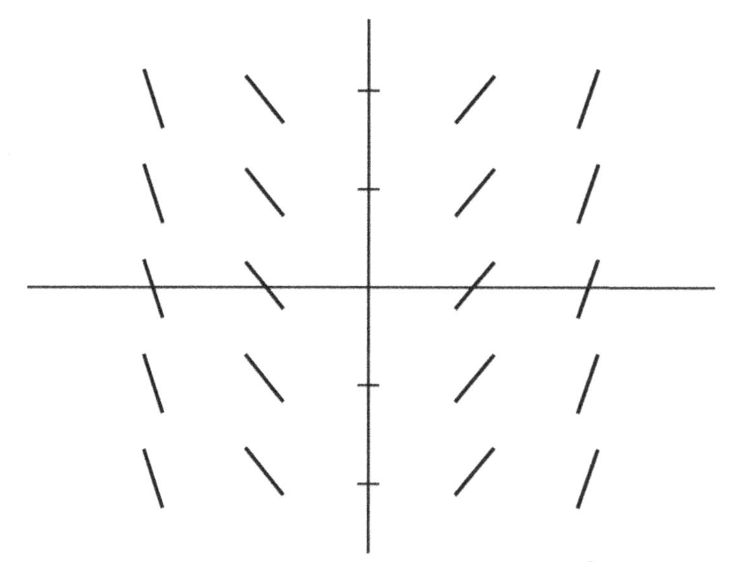

The slope field shown above represents the differential equation $\frac{dy}{dx} = 2x$. The solutions to this equation are the functions $y = x^2 + C$, as can be seen in the shapes formed by the slopes shown.

Separation of Variables

A certain class of differential equations, called separable equations, can be solved using antidifferentiation. The technique requires separating the variables so that each is represented only on a single side of the equation. Integrating both sides of the equation then produces a general solution. If an initial condition is provided, it can be used to find a particular solution. When integrating, it is only necessary to include a constant C on one side of the equation.

Example

Consider the differential equation $\frac{dy}{dt} = 6y^2t$ with initial condition $y(1) = 1$. To solve this, we begin by separating the variables: $\frac{1}{y^2}dy = 6t\,dt$. Integrating both sides, we have:

$$\int \frac{1}{y^2}dy = \int 6t\,dt \Rightarrow -\frac{1}{y} = 3t^2 + C.$$

This provides a general solution, albeit with an implicit definition. It is often recommended to solve for C using the initial condition before making y explicit. When the initial values are substituted, the result is C = –4. By utilizing this value and solving for y, we can derive the explicit solution:

$$-\frac{1}{y} = 3t^2 - 4$$

$$-y = \frac{1}{3t^2 - 4}$$

$$y = \frac{-1}{3t^2 - 4}$$

Exponential Models

Many applications of differential equations involve an exponential growth or decay model. This model occurs in any situation in which the rate of change of a quantity is proportional to the quantity. As an equation, this is represented by $\frac{dy}{dt} = ky$. This equation is easily solved using separation of variables, and the general solution is $y = y_0e^{kt}$, where y_0 is the value of y when $t =$ 0.

Example

The growth rate of a bacteria culture is directly related to the number of bacteria present. There is a particular scenario where the initial number of bacteria is 200, and after a span of 2 hours, the count increases to 1000. Let's discover the following:

a) The number of bacteria present after 5 hours
b) The time it will take for the culture to reach 7000 bacteria

To begin, note that since this follows exponential growth with a starting value of 200, the population is modeled by the equation $y = 200e^{kt}$. To solve for *k*, use the fact that after 2 hours there are 1000 bacteria:

$$\begin{aligned} 1000 &= 200e^{2t} \\ 5 &= e^{2t} \\ t &= \frac{\ln 5}{2} \end{aligned}$$

We can now rewrite the model as $y = 200e^{\frac{\ln 5}{2}t}$. The population after 5 hours is $200e^{\frac{5\ln 5}{2}} \approx 11180$.

For the second part of the problem, substitute $y = 7000$ in the model we found.

$$\begin{aligned} 7000 &= 200e^{\frac{\ln 5}{2}t} \\ 35 &= e^{\frac{\ln 5}{2}t} \\ \ln 35 &= \frac{\ln 5}{2}t \\ t &= \frac{2\ln 35}{\ln 5} \\ t &\approx 4.42 \end{aligned}$$

It will take approximately 4.42 hours for the population to reach 7,000.

Practice Differential Equations Questions

Do not use a calculator for the following two problems.

Which of the following is a solution of the first–order differential equation $y'(x) = 3y(x)$ subject to $y\left(\frac{1}{9}\right) = -1$?

A. $y(x) = -e^{\frac{1}{3}x - \frac{1}{27}}$

B. $y(x) = -e^{3x - \frac{1}{3}}$

C. $y(x) = e^{3x} - \sqrt[3]{e}$

D. $y(x) = -e^{3x}$

The correct answer is B. Verify that the function is a solution by plugging it into the differential equation and showing that it passes through the given point:

$$y(x) = -e^{3x - \frac{1}{3}} \Leftrightarrow y'(x) = -3e^{3x - \frac{1}{3}} = -3y(x)$$

$$y\left(\tfrac{1}{9}\right) = -e^{3\left(\frac{1}{9}\right) - \frac{1}{3}} = -e^{0} = -1$$

The general solution of $y'(x) = ky(x)$ is $y(x) = Ce^{kx}$, not $y(x) = Ce^{\frac{1}{k}x}$.

What is the general solution of the first-order differential equation $e^{3x}\dfrac{dy}{dx} = y^2$? (Note: In all choices, C represents an arbitrary real number.)

A. $y = \dfrac{e^{3x}}{1 + Ce^{3x}}$

B. $y = \dfrac{3e^{3x}}{1 + Ce^{3x}}$

C. $y = \sqrt[3]{e^{3x} + C}$

D. $y - 3e^{3x} + C$

The correct answer is B. This choice is the correct answer. Use separation of variables:

$$e^{3x}\frac{dy}{dx}=y^2$$
$$e^{3x}dy=y^2dx$$
$$y^{-2}dy=e^{-3x}dx$$
$$\int y^{-2}dy=\int e^{-3x}dx$$
$$-y^{-1}=-\tfrac{1}{3}e^{-3x}+C$$
$$\tfrac{1}{y}=\tfrac{1}{3}e^{-3x}+C$$
$$y=\frac{1}{\tfrac{1}{3}e^{-3x}+C}$$

Now, simplify as follows, recognizing that since C is any real number, it implies that 3C is also any real number. Therefore, it is acceptable to substitute C for 3C.

$$y=\frac{1}{\tfrac{1}{3}e^{-3x}+C}=\frac{1}{\frac{1}{3e^{3x}}+C}=\frac{1}{\frac{1}{3e^{3x}}+\frac{Ce^{3x}}{3e^{3x}}}=\frac{1}{\frac{1+Ce^{3x}}{3e^{3x}}}=\frac{3e^{3x}}{1+Ce^{3x}}$$

You may use a graphing calculator to solve the following problem.

The half-life of a specific radioactive isotope is 22.1 years. The starting quantity of the mass is 130 mg. Calculate the estimated duration for only 45 mg of the substance to remain.

A. 14.4 years
B. 33.8 years
C. 44.2 years
D. 46.9 years

The correct answer is B. A differential equation of the form

$$\begin{cases}\frac{dy}{dt}=ky\\ y(0)=130\end{cases}$$

where the constant k must be determined. The solution of this initial-value problem is $y(t)=130e^{kt}$.

Use the fact that $y(22.1)=65$ (half–life) to find k:

$$65 = 130e^{22.1k} \Rightarrow \tfrac{1}{2} = e^{22.1k}$$
$$\Rightarrow \ln\left(\tfrac{1}{2}\right) = 22.1k$$
$$\Rightarrow k = \tfrac{1}{22.1}\ln\left(\tfrac{1}{2}\right)$$

Therefore, the solution is $y(t) = 130e^{\left(\frac{1}{22.1}\ln\left(\frac{1}{2}\right)\right)t}$; this represents the amount of the substance (measured in mg) present at time t. We must determine the value of $t = t_0$ for which $y(t_0) = 45$. Use the graphing calculator to obtain the approximate solution graphically:

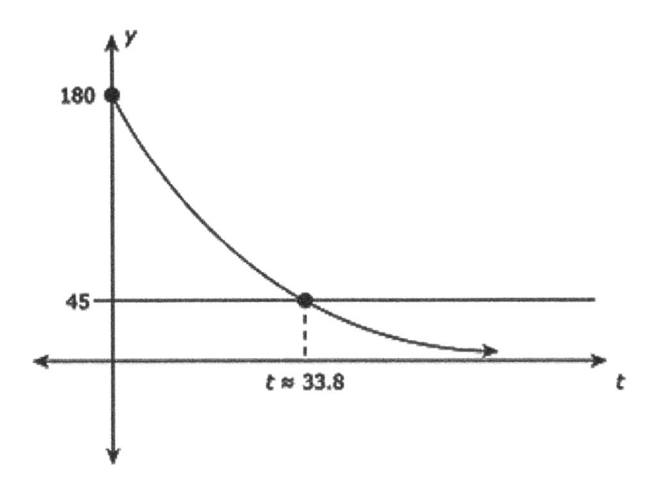

So, the approximate time it takes in order for only 45 mg of the substance to remain is 33.8 years.

Applications of Integration

Lastly, approximately 10–15% of the exam questions will focus on the Applications of Integration.

Average Value

If f is continuous on the interval $[a,b]$, then the average value of f on that interval is $\frac{1}{b-a}\int_a^b f(x)\,dx$. If f is nonnegative on the interval, the average value of the function has a simple graphical interpretation: it is the height a rectangle over the interval would have to be to have the same area as exists between the x-axis and the function.

Position, Velocity, and Acceleration

When a particle is moving along a straight line, its motion can be represented using derivatives, as we have previously discussed. Now, with the advancements in integration theory and techniques, we can expand upon this description by considering the following two points:

- The displacement of the particle over the time interval $[t_1, t_2]$ is given by $\int_{t_1}^{t_2} v(t)\,dt$, where $v(t)$ is the velocity of the particle.
- The total distance traveled by the particle over the time interval $[t_1, t_2]$ is $\int_{t_1}^{t_2} |v(t)|\,dt$. Recall that $|v(t)|$ is the speed of the particle at time t.

Accumulation Functions in Context

Integrating the rate of change allows us to determine the net change of a quantity over a given interval. This fact has significant implications across various applications.

Example

A tank of water contains 53 gallons at 8:00 AM. Between 8:00 AM and 12:00 PM, water leaks from the tank at a rate of $L(t) = 3|\sin t|$, where t is the number of hours since 8:00 AM, and L is measured in gallons per hour. How much water is remaining in the tank at 12:00 PM?

To solve this, we need to consider two quantities: the amount of water that the tank has at 8:00 AM, and the total amount of water that leaks from the tank between the hours of 8:00 AM and 12:00 PM. The first quantity is given as 53. The second quantity is the accumulation of the rate of leaking over the four hours. Therefore, the amount of water remaining in the tank at 12:00 PM is $53 - \int_0^4 3|\sin t|\,dt \approx 46$ gallons.

Free Response Tip

Pay attention to units in free response questions, as they are often required to be correct to receive full credit. Remember that the units of $\frac{dy}{dx}$ are the units of y over the units of x, and the units of $\int_a^b f(x)dx$ are the units of y times the units of x. When the units of y are a rate of change over time, and the units of x represent time, the units of the integral end up being equivalent to whatever quantity is changing.

Area Between Curves

If $f(x) \geq g(x)$ on the interval $[a,b]$, then the area between f on g between a and b is $\int_a^b [f(x) - g(x)]\,dx$. If the two functions intersect on an interval, the integral needs to be split into multiple subintervals, so that along each section the functions can be subtracted in the proper order.

Example

Let us find the area in the first quadrant bound by the graphs of $f(x) = \frac{1}{2}x$, $g(x) = x^2$, and the line $x = 1$. This is represented in the following graph.

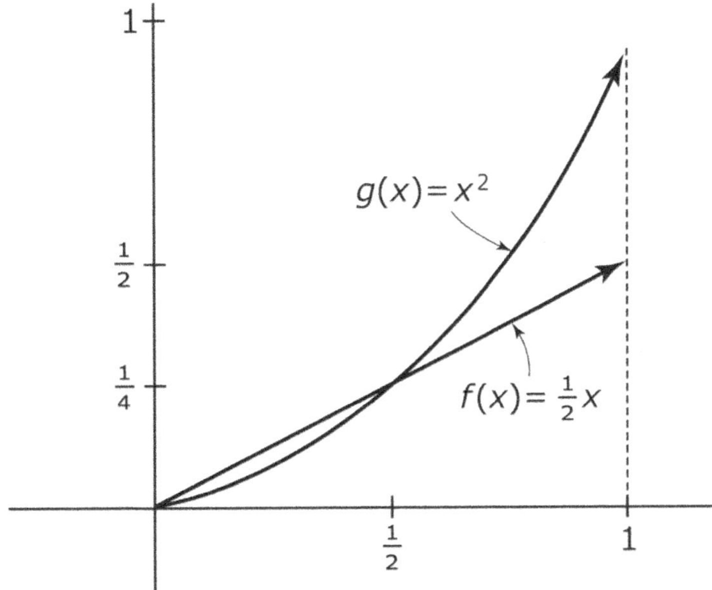

The graphs cross at $x = \frac{1}{2}$, so we will need to evaluate the two regions separately. The first region has area $\int_0^{1/2} \left(\frac{1}{2}x - x^2 \right) dx = \frac{1}{48}$, and the second region has area $\int_{1/2}^{1} \left(x^2 - \frac{1}{2}x \right) dx = \frac{5}{48}$.

The total area is $\frac{1}{48} + \frac{5}{48} = \frac{1}{8}$.

When curves are expressed in terms of y instead of x, the area between them can be determined using the same method. Instead of subtracting the function on the bottom from the function on the top, you should subtract the function on the left from the function on the right.

Volumes with Cross Sections

When a solid can be described in terms of its base and cross-sectional shapes, the volume of the solid can be computed by integrating the area of the cross sections along an appropriate interval. If the cross sections described are perpendicular to the x-axis, the volume is given by $V=\int_a^b A(x)\,dx$, where $A(x)$ is the area of the cross section in terms of x. If the cross sections are perpendicular to the y-axis, the integral is with respect to y: $V=\int_a^b A(y)\,dy$.

Shapes commonly used as cross sections include squares, rectangles, right triangles, equilateral triangles, and semicircles.

Example

The base of a solid is the region in the first quadrant of the xy-plane bounded by $y=\sqrt{x}$ and the vertical line $x=4$, shown as follows. Cross sections of the solid taken perpendicular to the y-axis are semicircles with diameter lying in the region given.

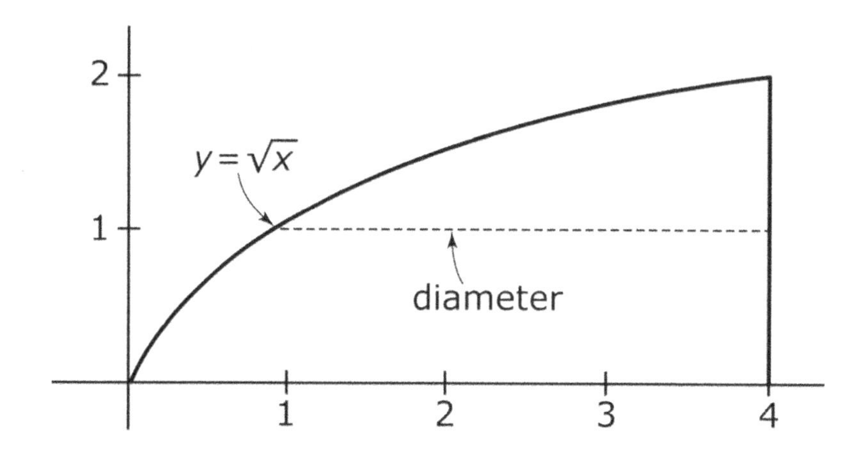

To find the volume of this solid, we need to first find a formula for the area of each cross section and appropriate limits of integration.

Since the cross sections are perpendicular to the y-axis, the diameter of each semicircle is the horizontal distance between $y=\sqrt{x}$ and $x=2$. Solving the square root function for x, we see that this distance is $4-y^2$. The radius of the semicircle, which we need to calculate its area, is half of this, or $r=\frac{1}{2}\left(4-y^2\right)=2-\frac{1}{2}y^2$. The area of the semicircle is

$$A(y)=\frac{1}{2}\pi r^2=\frac{1}{2}\pi\left(2-\frac{1}{2}y^2\right)^2=\pi\left(2-y^2+\frac{1}{8}y^4\right).$$

The limits of integration are the range of y-values that span the region. The lower boundary is $y = 0$, and the upper boundary is $y = 2$. Therefore, the volume of the region is

$$V = \pi\int_0^2\left(2 - y^2 + \frac{1}{8}y^4\right)dy = \frac{32\pi}{15}.$$

Free Response Tip

Free response volume questions are commonly found on the section that prohibits the use of a calculator. Typically, the questions are presented in the following manner: "Create an integral that represents the given situation, but do not calculate its value." It is advised to refrain from attempting to calculate a numerical solution. Instead, it is best to consider the integral as the final answer.

Solids of Revolution

When a solid has circular or washer (ring-shaped) cross sections perpendicular to a vertical or horizontal line, it can be described as being obtained by rotating a region around that vertical or horizontal line. For this situation, the volume can be determined using a standard formula:

- If the axis of revolution is horizontal, and the cross sections are circles, the volume is $V = \pi\int_a^b r^2\,dx$, where r is the radius in terms of x.
- If the axis of revolution is horizontal, and the cross sections are washers, the volume is $V = \pi\int_a^b\left(R^2 - r^2\right)dx$, where R is the radius of the outer circle, and r is the radius of the inner circle.
- If the axis of revolution is vertical, and the cross sections are circles, the volume is $V = \pi\int_a^b r^2\,dy$, where r is the radius in terms of y.
- If the axis of revolution is vertical, and the cross sections are washers, the volume is $V = \pi\int_a^b\left(R^2 - r^2\right)dy$, where R is the radius of the outer circle, and r is the radius of the inner circle.

> **Free Response Tip**
>
> When solving volume problems that permit calculator use, it is important to not only provide the calculated answer but also demonstrate the integral being evaluated. Simply providing the numerical answer will not be enough to receive full credit.

Example

The region bound by the graphs of $y = x^2$ and $y = \sqrt{x}$ is revolved around the line $y = 2$. The cross sections perpendicular to the x-axis are washers, with $R = 2 - x^2$, and $r = 2 - \sqrt{x}$, shown as follows.

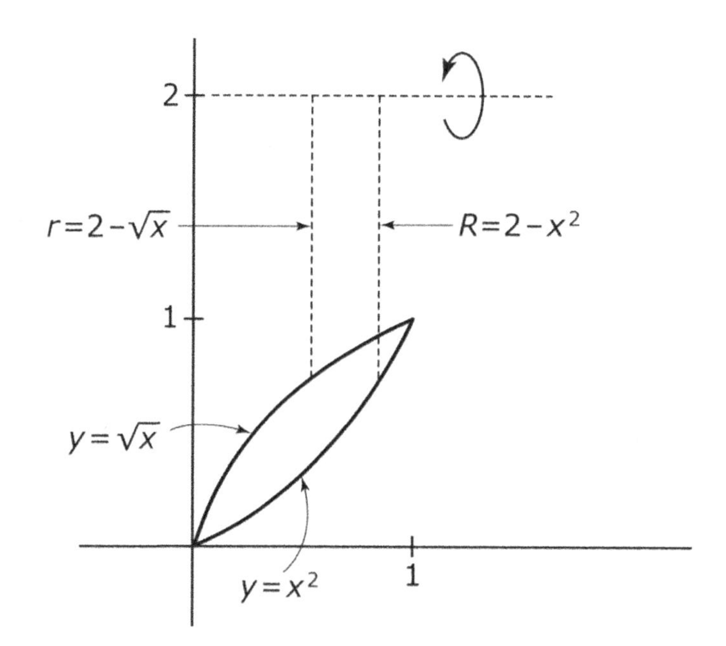

The volume of the solid is $V = \pi\int_0^1 \left[\left(2 - x^2\right)^2 - \left(2 - \sqrt{x}\right)^2\right] dx = \frac{31\pi}{30}$.

Sample Applications of Integration Questions

Do not use a calculator for the following two problems.

Suppose a chemical leaks from a holding tank into a reservoir at a rate of $R(t)$ gallons per minute at time t. Which of the following expressions equals the number of gallons of chemical that have leaked from the tank during the first three hours?

A. $\int_0^{180} R'(t)dt$

B. $\int_0^{180} R(t)dt$

C. $\int_0^{3} R(t)dt$

D. $R(180) - R(0)$

The correct answer is B. $R(t)$ is already the rate at which the chemical leaks from the tank and its units are gallons per minute. Since time is measured in minutes, the "first three hours" corresponds to the time interval [0, 180] minutes; this is the interval of integration. So, by the net change theorem, the number of gallons of chemical that have leaked from the tank during the first three hours is given by $\int_0^{180} R(t)dt$.

Suppose the graph of the function $m(x)$ on the interval $[-5,-1]$ is as follows:

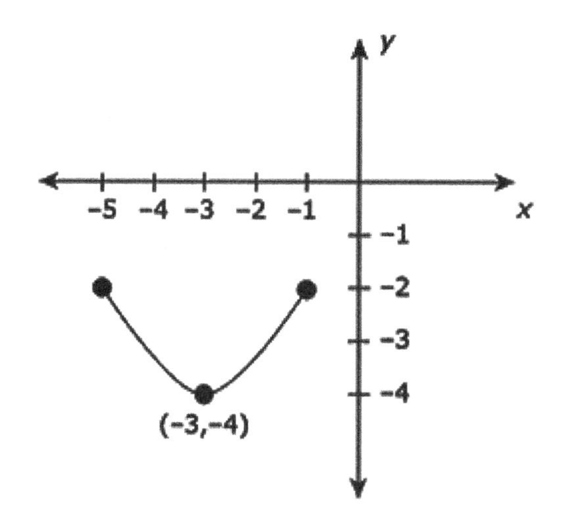

(Assume the curve forms a semi–circle.) Compute $\int_{-5}^{-1} m(x)dx$.

A. $-8-2\pi$
B. -2π
C. 2π
D. $8+2\pi$

The correct answer is A. Divide the region bounded by the curve $y = m(x)$ and the vertical lines $x = -5$ and $x = -1$ into two pieces: a rectangle and a semicircle of radius 2. The area of the rectangle is $4(2) = 8$ and the area of the semicircle is $\frac{1}{2}\cdot\pi(2)^2 = 2\pi$. Since both regions are below the x-axis, they contribute a negative amount to the integral $\int_{-5}^{-1} m(x)dx$. Hence, $\int_{-5}^{-1} m(x)dx = -8-2\pi$.

You may use a graphing calculator to solve the following problem.

Consider the function $g(x) = e^{\sin x} - 1$ on the interval $[0, 2\pi]$. What is the smallest positive approximate value of c in this interval for which $g(c)$ *equals* the average of $g(x)$ on $[0, 2\pi]$?

A. 0.238
B. 1.385
C. 1.571
D. 2.903

The correct answer is A. By definition, the average value of $g(x) = e^{\sin x} - 1$ on $[0, 2\pi]$ is given by $\frac{1}{2\pi}\int_0^{2\pi}\left(e^{\sin x} - 1\right)dx$. Using the graphing calculator, we see that this quantity is approximately equal to 0.26607. Now, we must find the points of intersection of $g(x)=e^{\sin x}-1$ and the horizontal line $y=0.26607$ in the interval $[0,2\pi]$:

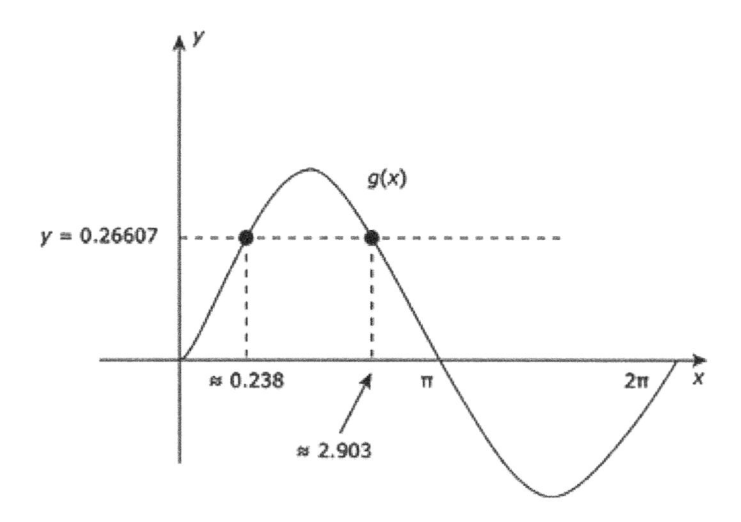

Of the two such points of intersection, 0.238 is the smallest value of x. This is the value we seek.

Practice Exam

Non-Calculator Multiple Choice

1) A particle moves along a straight line so that at time t $\geq$ 0 its acceleration is given by the function $a(t) = 4t$. At time t = 0, the velocity of the particle is 4 and the position of the particle is 1. Which of the following is an expression for the position of the particle at time t $\geq$ 0?

(a) $\frac{2}{3}t^3 + 4t + 1$

(b) $2t^3 + 4t + 1$

(c) $\frac{1}{3}t^3 + 4t + 1$

(d) $\frac{2}{3}t^3 + 4$

2)

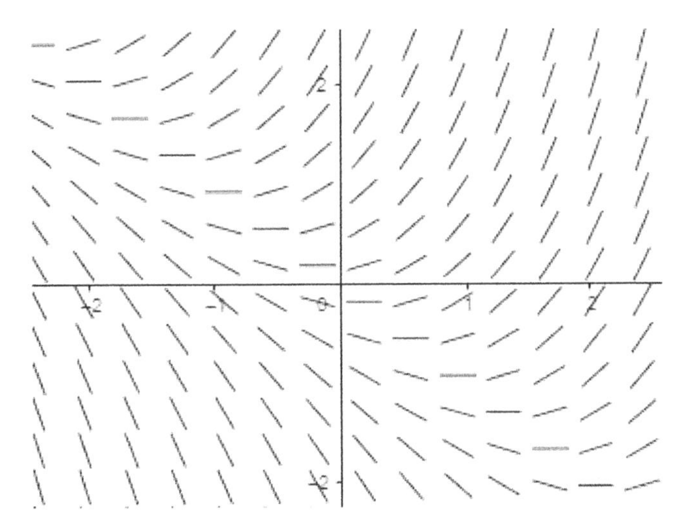

Shown above is a slope field for which of the following differential equations?

(a) $\frac{dy}{dx} = \frac{x}{y}$ (b) $\frac{dy}{dx} = xy$ (c) $\frac{dy}{dx} = x + y$ (d) $\frac{dy}{dx} = x - y$

3)

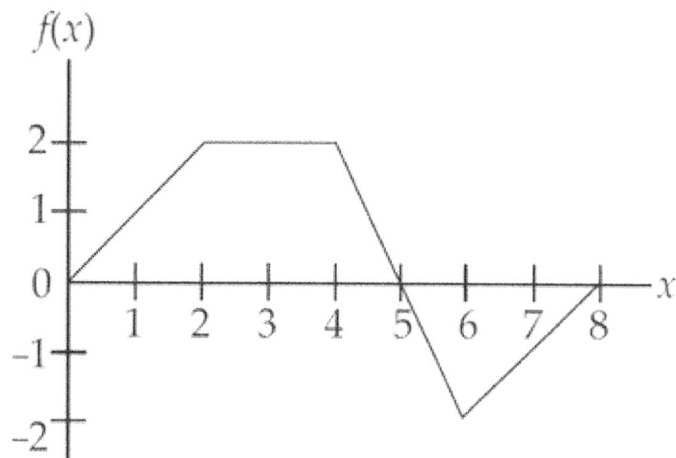

The graph of a piecewise linear function f(x) is above. Evaluate $\int_3^8 f'(x)\,dx$

(a) 2 (b) -2 (c) 5 (d) 0

4)

$$\int_1^5 \frac{x-1}{x}\,dx$$

(a) $5 - \ln 5$ (b) $4 - \ln 5$ (c) $2 - \ln 5$ (d) $1 - \ln 5$

5)

$\lim_{x \to 1} \dfrac{2 \cdot \ln(x)}{e^x - 1}$ is

(a) $\frac{2}{e}$ (b) 1 (c) 0 (d) *nonexistent*

$$f(x) = \begin{cases} x + 5 & x < -2 \\ x^2 + 2x + 3 & x \geq -2 \end{cases}$$

6) Let f be the function defined above. Which of the following statements about f is true?

(a) f is continuous and differentiable at x = -2.

(b) f is continuous but not differentiable at x = -2.

(c) f is differentiable but not continuous at x = -2.

(d) f is defined but is neither continuous nor differentiable at x = -2.

7) The equation $y = e^{2x}$ is a particular solution to which of the following differential equations?

(a) $\frac{dy}{dx} = 1$ (b) $\frac{dy}{dx} = y$ (c) $\frac{dy}{dx} = y + 1$ (d) $\frac{dy}{dx} = y - 1$

8) For any real number x, $\lim_{h \to 0} \frac{cos((x+h)^2) - cos(x^2)}{h} =$

(a) $cos(x^2)$ (b) $2xcos(x^2)$ (c) $-sin(x^2)$ (d) $-2xsin(x^2)$

9) What is the value of x at which the maximum value of $y = \frac{4}{3}x^3 - 8x^2 + 15x$ occurs on the closed interval [0, 4]?

(a) 0 (b) $\frac{3}{2}$ (c) $\frac{5}{2}$ (d) 4

10) At time t = 0, a reservoir begins filling with water. For t > 0 hours, the depth of the water in the reservoir is increasing at a rate of R(t) inches per hour. Which of the following is the best interpretation of $R'(2) = 4$?

(a) The depth of the water is 4 inches, at t = 2 hours.

(b) The depth of the water is increasing at a rate of 4 inches per hour, at t = 2 hours.

(c) The rate of change of the depth of the water is increasing at a rate of 4 inches per hour per hour at t = 2 hours

(d) The depth of the water increased by 4 inches from t = 0 to t = 2 hours.

11) If $f'(x) = 3x^2$ and $f(2) = 3$, then $f(1) =$

(a) -7 (b) -4 (c) 7 (d) 10

12) A particle moves along the x-axis so that at time $t \geq 0$ its velocity is given by $v(t) = e^{t-1} - 3sin(t-1)$. Which of the following statements describes the motion of the particle at time $t = 1$?

(a) The particle is speeding up at $t = 1$.

(b) The particle is slowing down at $t = 1$.

(c) The particle is neither speeding up nor slowing down at $t = 1$.

(d) The particle is at rest at $t = 1$.

13)

x	0	2	4	6	8
$f(x)$	1	-1	-5	7	5

The table above gives selected values for the twice-differentiable function f. In which of the following intervals must there be a number c such that $f'(c) = -2$.

(a) $(0, 2)$ (b) $(2, 4)$ (c) $(4, 6)$ (d) $(6, 8)$

14) $\frac{d}{dx}(tan(ln(x))) =$

(a) $\frac{tan(ln(x))}{x}$ (b) $sec^2(ln(x))$ (c) $\frac{sec^2(ln(x))}{x}$ (d) $tan(\frac{1}{x})$

15) The function f is given by $f(x) = x^3 - 2x^2$. On what interval(s) is f(x) concave down?

(a) $(0, \frac{4}{3})$ (b) $(-\infty, 0)$ and $(\frac{4}{3}, \infty)$ (c) $(-\infty, \frac{2}{3})$ (d) $(\frac{2}{3}, \infty)$

16) If $\sqrt{x} + y^2 = xy + 2$, what is $\frac{dy}{dx}$ at the point (4,0)?

(a) $-\frac{1}{16}$ (b) $\frac{1}{16}$ (c) $-\frac{1}{4}$ (d) $\frac{1}{4}$

17) Let R be the region bounded by the graphs of $y = 2x$ and $y = x^2$. What is the area of R?

(a) 0 (b) 4 (c) $\frac{2}{3}$ (d) $\frac{4}{3}$

18) A block of ice in the shape of a cube melts uniformly maintaining its shape. The volume of a cube given a side length is given by the formula $V = S^3$. At the moment S = 2 inches, the volume of the cube is decreasing at a rate of 5 cubic inches per minute. What is the rate of change of the side length of the cube with respect to time, in inches per minute, at the moment when S = 2 inches?

(a) $-\frac{5}{12}$ (b) $\frac{5}{12}$ (c) $-\frac{12}{5}$ (d) $\frac{12}{5}$

19) Let g be the function given by $f(x) = \int_1^x (3t - 6t^2)dt$ .What is the x-coordinate of the point of inflection of the graph of f?

(a) $-\frac{1}{4}$ (b) $\frac{1}{4}$ (c) 0 (d) $\frac{1}{2}$

20) $\int sin(3x)dx =$

(a) $3cos(3x) + C$

(b) $\frac{1}{3}cos(3x) + C$

(c) $-3cos(3x) + C$

(d) $-\frac{1}{3}cos(3x) + C$

21) How many removable discontinuities does the graph of $y = \frac{x-2}{x^4-16}$ have?

(a) one (b) two (c) three (d) four

22) If $\int_4^6 f(x)dx = 5$ and $\int_{10}^4 f(x)dx = 8$ then what is the value of $\int_6^{10} (4f(x) + 10)dx$

(a) -12 (b) 12 (c) 52 (d) 62

23) What is the equation of the line tangent to the graph $y = e^{2x}$ at $x = 1$?

(a) $y + 2e^2 = e^2(x-1)$

(b) $y + e^2 = 2e^2(x-1)$

(c) $y - 2e^2 = e^2(x-1)$

(d) $y - e^2 = 2e^2(x-1)$

24)

$$\lim_{x \to \infty} \frac{4 \cdot \ln(x) + 4}{3x} =$$

(a) 2 (b) -2 (c) 0 (d) *nonexistent*

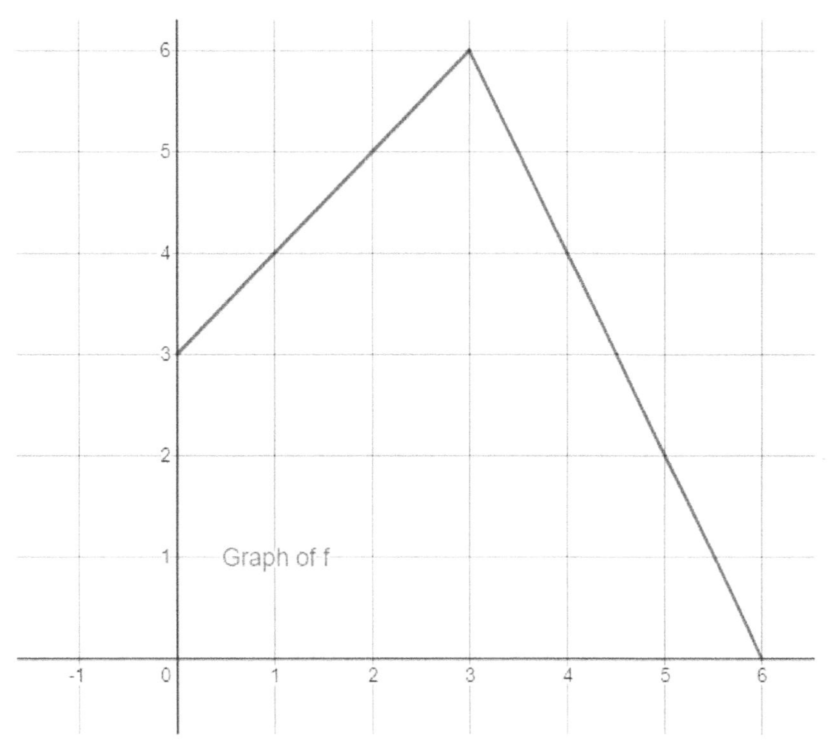

25) The graph of a function, f is shown above. Let h(x) be defined as $h(x) = (x+1) \cdot f(x)$. Find $h'(4)$.

(a) -6 (b) -2 (c) 4 (d) 14

26) A region R is the base of a solid where $f(x) \geq g(x)$ for all x $a \leq x \leq b$. For this solid, each cross section perpendicular to the x-axis are rectangles with height 5 times the base. Which of the following integrals gives the volume of this solid?

(a) $25\int_a^b (g(x) - f(x))^2 dx$

(b) $5\int_a^b (g(x) - f(x))dx$

(c) $5\int_a^b (f(x) - g(x))dx$

(d) $5\int_a^b (f(x) - g(x))^2 dx$

27) If $\frac{dy}{dx} = \frac{x}{y}$ and if y = 4 when x = 2, then y =

(a) $\sqrt{\frac{1}{2}x^2 + 14}$

(b) $\sqrt{2x^2 + 8}$

(c) $\sqrt{x^2 + 6}$

(d) $\sqrt{x^2 + 12}$

28) If $f(x) = \frac{x^2 + 1}{3x}$ then $f'(x) =$

(a) $\frac{3 - 3x^2}{9x^2}$

(b) $\frac{3x^2 - 3}{9x^2}$

(c) $\frac{3x^2 + 3}{9x^2}$

(d) $\frac{2x}{3}$

29) $\int (2x + 3)(x^2 + 3x)^4 dx =$

(a) $\frac{1}{5}(x^2 + 3x)^5 + C$

(b) $\frac{1}{10}(x^2 + 3x)^5 + C$

(c) $(x^2 + 3x)^5 + C$

(d) $5(x^2 + 3x)^5 + C$

x	1	4	6	7
f(x)	3	5	2	8
g(x)	2	1	0	5

30) Two differentiable functions, f and g have the property that $f(x) \geq g(x)$ for all real numbers and form a closed region R that is bounded from x = 1 to x = 7. Selected values of f and g are in the table above. Estimate the area between the curves f and g between x = 1 and x = 7 using a Right Riemann sum with the three sub-intervals given in the table.

(a) 13 (b) 19 (c) 21 (d) 27

Calculator Active Multiple Choice

$$f(x) = \begin{cases} -2x^2 - 7 & \text{if } x \leq 1 \\ -x^2 - 2kx & \text{if } x > 1 \end{cases}$$

76) Let f be the function defined above, where k is a constant. For what value of k, is f(x) continuous at x = 1?

(a) $-\frac{9}{2}$ (b) -4 (c) 4 (d) $\frac{9}{2}$

77) At time t, $0 < t < 2$, the velocity of a particle moving along the x-axis is given by $v(t) = e^{t^2} - 2$. What is the total distance traveled by the particle during the time interval $0 < t < 2$?

a) 12.453 (b) 13.368 (c) 51.598 (d) 53.598

78) Let f be a continuous function such that $\int_{2}^{5} f(x)\,dx = -4$ and $\int_{8}^{5} f(x)\,dx = 3$

then $\int_{8}^{2} f(x)\,dx =$

(a) -7 (b) -1 (c) 1 (d) 7

79) Let f be a twice-differentiable function defined by the differentiable function g such that $f(x) = \int_{-2}^{x} g(x)\,dx$. It is also known that g(x) is always concave up, decreasing, and positive for all real numbers. Which of the following could be false about f(x)?

(a) f(x) is concave down for all x

(b) f(x) is increasing for all x

(c) f(x) is negative for all x

(d) f(x) = 0 for some x in the real numbers

80) Let f be the function defined by $f(x) = e^{cos(x)} - sin(x)$. For what value of x, on the interval (0,4), is the average rate of change of f(x) equal to the instantaneous rate of change of f(x) on [0,4]?

(a) 0.723 (b) 1.901 (c) 1.966 (d) 2.110

81) Let f and g be differentiable functions such that $f(g(x)) = x$ for all x. If $f(1) = 3$ and $f'(1) = -4$, what is the value of g'(3)?

(a) $\frac{1}{3}$ (b) $-\frac{1}{3}$ (c) $\frac{1}{4}$ (d) $-\frac{1}{4}$

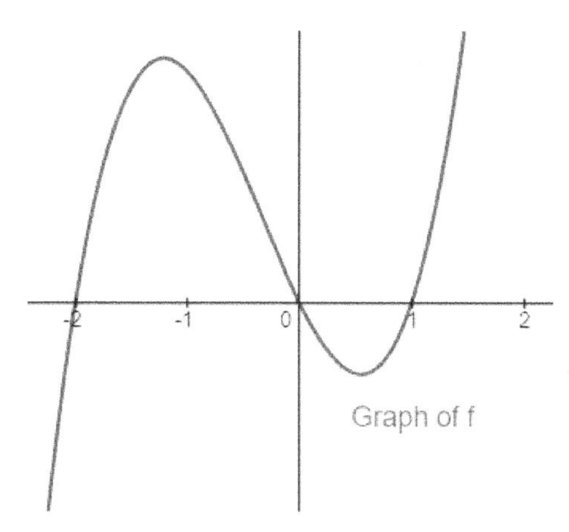

82) The graph of $y = f(x)$ is shown above. Which of the following could be the graph of $y = f'(x)$?

(a)

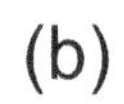

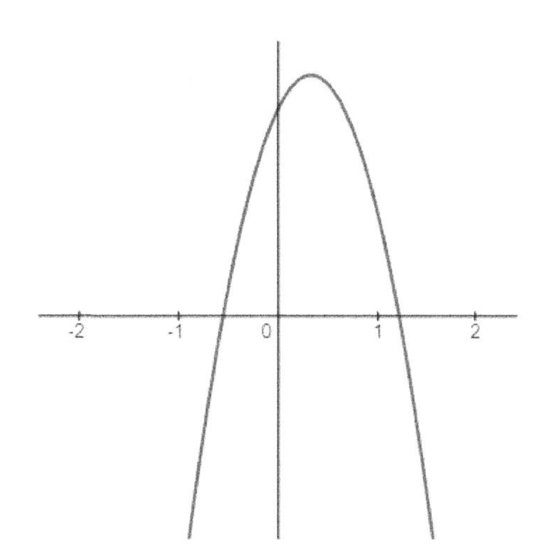

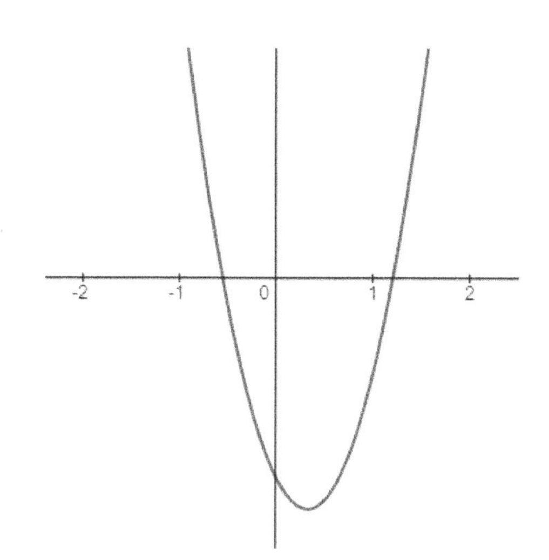

(c)

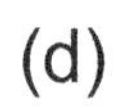

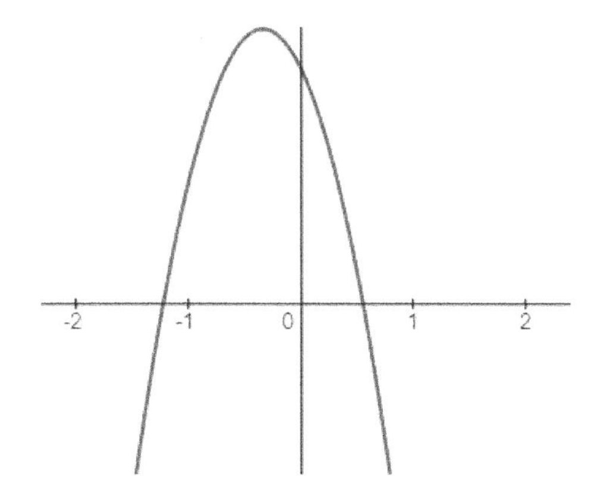

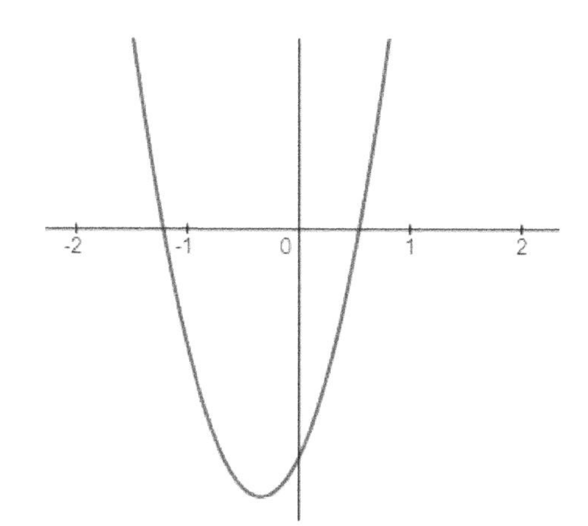

x	-5	0	3
f(x)	6	4	-2

83) The table above gives values of a differentiable function f(x) at selected x values. Based on the table, which of the following statements about f(x) could be false?

(A) There exists a value c, where $-5 < c < 3$ such that $f(c) = 1$

(B) There exists a value c, where $-5 < c < 3$ such that $f'(c) = 1$

(C) There exists a value c, where $-5 < c < 3$ such that $f(c) = -1$

(D) There exists a value c, where $-5 < c < 3$ such that $f'(c) = -1$

84) The function f is the antiderivative of the function g defined by $g(x) = e^x - ln(x) - 2x^2$. Which of the following is the x-coordinate of location of a relative maximum for the graph of $y = f(x)$.

(a) 1.312 (b) 2.242 (c) 2.851 (d) 2.970

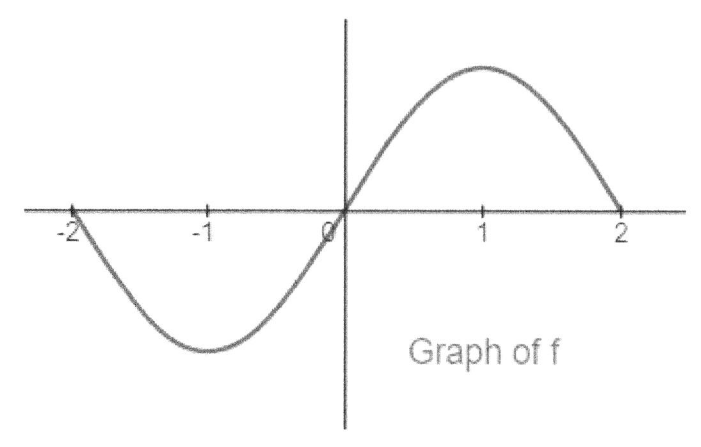

85) The function f is continuous on the closed interval [-2,2]. The graph of f′, the derivative of f, is shown above. On which interval(s) is f(x) increasing?

(a) [-1,1]

(b) [-2, -1] and [1, 2]

(c) [0, 2]

(d) [-2, 0]

86) Let f be the function with the first derivative $f'(x) = \sqrt{sin(x) + cos(x) + 2}$. If g(3) = 4, what is the value of g(6)?

(a) 2.328 (b) 3.918 (c) 6.328 (d) 7.918

87) The velocity of a particle for $t \geq 0$ is given by $v(t) = ln(t^3 + 1)$. What is the acceleration of the particle at t = 4 ?

(a) 0.738 (b) 3.436 (c) 4.174 (d) 8.232

88) The function f is differentiable and f(4) = 3 and f'(4) = 2. What is the approximation of f(4.1) using the tangent line to the graph of f at x = 4 ?

(a) 2.6 (b) 2.8 (c) 3.2 (d) 3.4

89) Patrick is climbing stairs and the rate of stair climbing is given by the differentiable function s, where s(t) is measured in stairs per second and t is measured in seconds. Which of the following expressions gives Patrick's average rate of stairs climbed from t = 0 to t = 20 seconds?

(a) $\int_0^{20} s(t)dt$

(b) $\frac{1}{20}\int_0^{20} s(t)dt$

(c) $\int_0^{20} s'(t)dt$

(d) $\frac{1}{20}\int_0^{20} s'(t)dt$

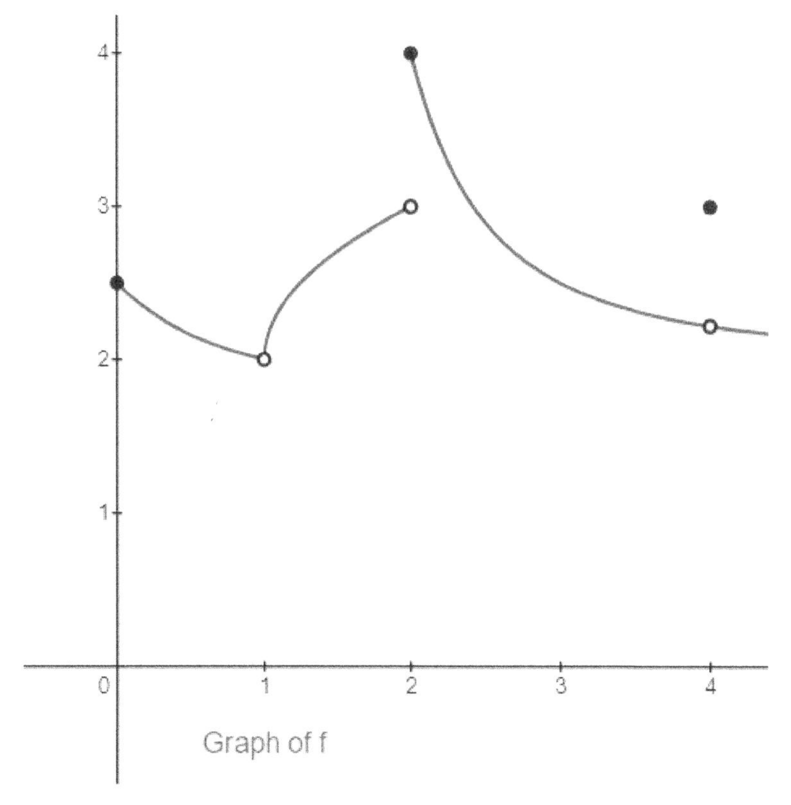

90) The graph of f is shown above. Which of the following statements is false?

(a) $f(1) = \lim_{x \to 1^-} f(x)$

(b) $f(3) = \lim_{x \to 3} f(x)$

(c) f(x) has a jump discontinuity at x = 2

(d) f(x) has a removable discontinuity at x = 4

Free Response Question with Calculator

1) A water bottle has a height of 18 centimeters and has circular cross sections. The radius, in centimeters, of a circular cross section of the bottle at height *h* centimeters is given by the piecewise function:

$$R(h) = \begin{cases} 3 & 0 \le h < 12 \\ 3 - \frac{1}{13}(h-12)^2 & 12 \le h \le 18 \end{cases}$$

(a) Is R(h) continuous at h = 12? Justify your response.

(b) Find the average value of the radius from h = 12 to h = 18.

(c) Find the volume of the water bottle. Include units.

(d) The water bottle is being filled up at a hydration station. At the instant when the height of the liquid is $h = 14$ centimeters, the height is increasing at a rate of $\frac{3}{4}$ centimeters per second. At this instant, what is the rate of change of the radius of the cross section of the liquid with respect to time?

2) Coal is burning in a furnace, thus exhausting the resource. The rate at which coal is burning, measured in pounds per hour, is given by $B(t) = 4sin(\frac{t}{2})$. At t = 2 hours, a worker starts supplying additional coal into the furnace. The rate at which coal is being added, measured in pounds per hour, is given by $S(t) = 12 + 10sin(\frac{4\pi t}{25})$. The worker stops adding coal at t = 6 hours. At t = 0 there are 500 pounds of coal in the furnace.

(a) Find the total amount of coal added by the worker. Include units.

(b) Is the amount of coal in the furnace increasing or decreasing at t = 5 hours? Explain.

(c) Find $S'(4)$. Explain, with units, the meaning of this in the context of the problem.

(d) Find the amount of coal, in pounds, in the furnace at $t = 6$ hours.

Free Response Question without Calculator

3)

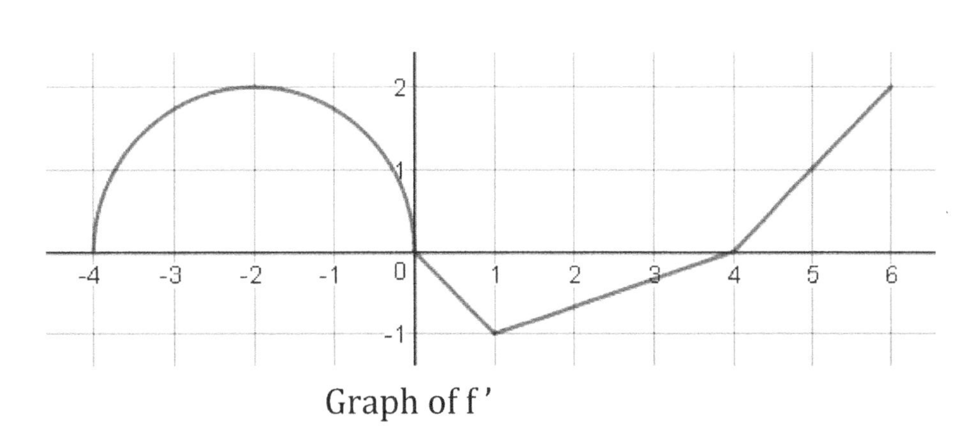

Graph of f '

The figure above represents the function f ' the derivative of f over the interval [-4, 6] and satisfies f(4) = 2. The graph of f ' consists of three line segments and a semi-circle.

(a) Find the value of f(-4).

(b) On what interval(s) is f decreasing and concave up? Justify your answer.

(c) State all x-values where f(x) has a horizontal tangent on the open interval (-4, 6). Explain whether f has a relative minimum, relative maximum, or neither at each of those x-values.

(d) Evaluate $\int_{2}^{3} f''(2x)dx$

4) Consider the differential equation, the derivative of f(x), $\frac{dy}{dx} = \frac{2y^2}{x-1}$ and where f(2) = 1.

(a) On the axes below, sketch a slope field for the given differential equation at the six points indicated.

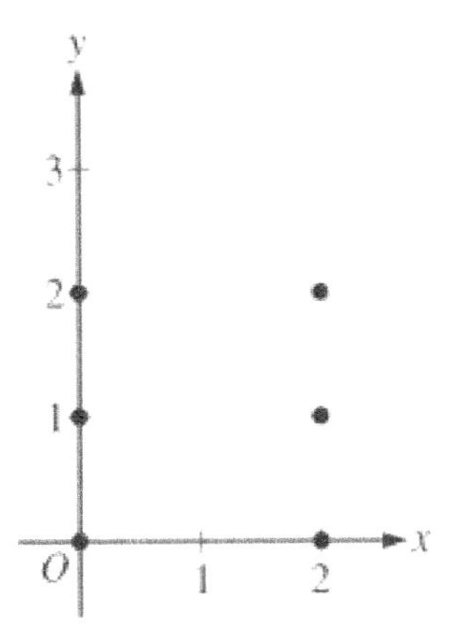

(b) Find an expression for f(x) given that f(2) = 1.

(c) Find $\frac{d^2y}{dx^2}$

x	1	4	6	9
f(x)	10	3	5	2

5) Let g(x) be a twice-differentiable function defined by a differentiable function f, such that $g(x) = 2x + \int_{1}^{x^2} f(t)dt$. Selected values of f(x) are given in the table above.

(a) Use a Left Riemann sum using the subintervals indicated by the table to approximate g(3).

(b) Find g'(x) and evaluate g'(3).

(c) Using the data in the table, estimate $f'(3)$.

(d) Explain why there must be a value of f, on $1 < x < 9$ such that $f(c) = 4$

6)

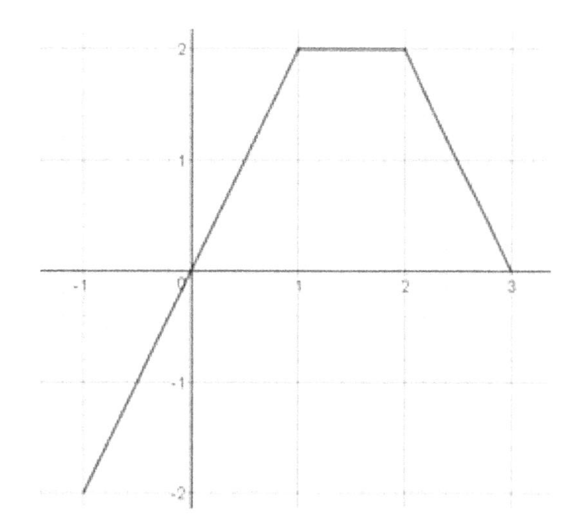

Graph of f(x)

x	g(x)	g'(x)
0	2	-4
1	1	-2
5/2	4	-3
5	-1	3

Let f be a continuous function defined on the interval [-1, 3], and whose graph is given above. Let g be a differentiable function with derivative g'. The table above gives the value of g(x) and g'(x) at selected x-values.

(a) Let h be the function defined by $h(x) = f(x) \cdot g(x)$. Find $h'(0)$

(b) Let k be the function defined by $k(x) = g(f(x))$. Find the slope of the line tangent to k(x) at $x = \frac{5}{2}$.

(c) Let $m(x) = \int_1^x g'(t)dt$. Find m(5) and m'(5)

(d) Evaluate $\lim_{x \to 0} \frac{g(x) - 2}{f(2x)}$

Multiple Choice Answers

1) A	24) C
2) C	25) A
3) B	26) D
4) B	27) D
5) A	28) B
6) B	29) A
7) C	30) B
8) D	
9) D	76) C
10) C	77) B
11) B	78) D
12) B	79) C
13) B	80) B
14) C	81) D
15) C	82) D
16) B	83) B
17) D	84) A
18) A	85) C
19) B	86) D
20) D	87) A
21) A	88) C
22) A	89) B
23) D	90) A

Free Response Questions Solutions

1) A water bottle has a height of 18 centimeters and has circular cross sections. The radius, in centimeters, of a circular cross section of the bottle at height h centimeters is given by the piecewise function:

$$R(h) = \begin{cases} 3 & 0 \le h < 12 \\ 3 - \frac{1}{13}(h-12)^2 & 12 \le h \le 18 \end{cases}$$

(a) Is R(h) continuous at h = 12? Justify your response.

$\lim_{h \to 12^-} R(h) = 3$ $\lim_{h \to 12^+} R(h) = R(12) = 3$

Since $\lim_{h \to 12^-} R(h) = \lim_{h \to 12^+} R(h) = R(12) = 3,$ R(h) is continuous at h = 12.

(b) Find the average value of the radius from h = 12 to h = 18.

$$\tfrac{1}{6}\int_{12}^{18} 3 - \tfrac{1}{13}(h-12)^2 dh = 2.077 \ (or\ 2.076)\ centimeters$$

(c) Find the volume of the water bottle. Include units.

$$\pi \int_0^{12} (3)^2 dh + \pi \int_{12}^{18} [3 - \tfrac{1}{13}(h-12)^2]^2 dh = 433.451\ cm^3\ (or\ 433.450)$$

(d) The water bottle is being filled up at a hydration station. At the instant when the height of the liquid is h = 14 centimeters, the height is increasing at a rate of $\frac{3}{4}$ centimeters per second. At this instant, what is the rate of change of the radius of the cross section of the liquid with respect to time?

$$R = 3 - \frac{1}{13}(h-12)^2$$

$$\frac{d}{dt}\left[R = 3 - \frac{1}{13}(h-12)^2\right]$$

$$\frac{dR}{dt} = -\frac{2}{13}(h-12) * \frac{dh}{dt}$$

$$\frac{dR}{dt} = -\frac{2}{13}(14-12) * \frac{3}{4}$$

$$\frac{dR}{dt} = -0.231\ or\ -0.230\ cm/sec$$

2) Coal is burning in a furnace, thus exhausting the resource. The rate at which coal is burning, measured

in pounds per hour, is given by $B(t) = 4sin(\frac{t}{2})$. At t = 2 hours a worker starts supplying additional coal into the furnace. The rate at which coal is being added, measured in pounds per hour, is given by $S(t) = 12 + 10sin(\frac{4\pi t}{25})$. The worker stops adding coal at t = 6 hours. At t = 0 there are 500 pounds of coal in the furnace.

(a) Find the total amount of coal added by the worker. Include units.

(a) Coal added = $\int_2^6 S(t)\,dt$ = 78.397 pounds

78.397 pounds of coal were added by the worker.

(b) Is the amount of coal in the furnace increasing or decreasing at t = 5 hours? Explain.

(b) $S(5) = 17.878 \; or \; 17.877 \frac{lbs}{hr}$

$B(5) = 2.394 \; or \; 2.393 \frac{lbs}{hr}$

$S(5) > B(5)$ therefore, the amount of coal in the furnace is increasing at t = 5 hours.

(c) Find S'(4). Explain, with units, the meaning of this in the context of the problem.

(c) S'(4) = -2.140 lbs/hr^2

S'(4) means the rate at which coal is being added to the furnace is decreasing at a rate of 2.140 lbs/hr²

(d) Find the amount of coal, in pounds, in the furnace at t = 6 hours.

(d) Let $A(t) = 500 + + \int_2^t S\left(x\right) dx - \int_0^t B(x)dx \; for \; t \geq 2$

A(6) = 500 + $\int_2^6 S(t)\,dt$ - $\int_0^6 B(t)\,dt$

= 562.477 pounds

There are 562.477 pounds of coal in the furnace at t = 6 hours.

Non-Calculator Free Response

3)

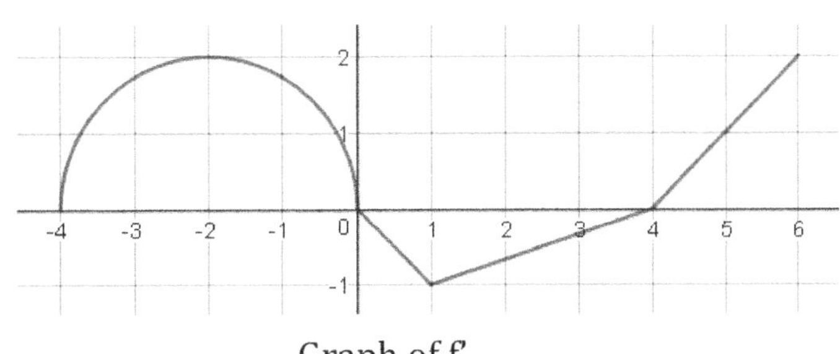

Graph of f'

The figure above represents the function f' a continuous function, the derivative of f over the interval [-4, 6] and satisfies f(0) = 4. The graph of f' consists of three line segments and a semi-circle.

(a) Find the value of f(-4).

$$f(-4) = 4 + \int_0^{-4} f'(x)\,dx = 4 - 2\pi$$

(b) On what interval(s) is f decreasing and concave up? Justify your answer.

(b) $f(x)$ is decreasing and concave up on the interval (1,4) because f'(x) is negative and increasing.

(c) State all x-values where f(x) has a horizontal tangent on the open interval (-4, 6). Explain whether f has a relative minimum, relative maximum, or neither at each of those x-values.

(c) At x = 0 and x =4, f(x) has a horizontal tangent.
At x = 0, f(x) has a relative maximum because f' changes from positive to negative.
At x = 4, f(x) has a relative minimum because f' changes from negative to positive.

(d) Evaluate $\int_2^3 f''(2x)dx$

$$\int_2^3 f''(2x)dx = \tfrac{1}{2}f'(2x)\Big|_2^3 = \tfrac{1}{2}f'(6) - \tfrac{1}{2}f'(4) = 1$$

4) Consider the differential equation, the derivative of f(x), $\frac{dy}{dx} = \frac{2y^2}{x-1}$ and where f(2) = 1.

(a) On the axes below, sketch a slope field for the given differential equation at the six points indicated.

(b) Find an expression for f(x) given that f(2) = 1.

$\frac{1}{2y^2}dy = \frac{1}{x-1}dx$

$\int \frac{1}{2y^2}dy = \int \frac{1}{x-1}dx$

$\frac{-1}{2y} = ln|x-1| + C$

$\frac{-1}{2(1)} = ln|2-1| + C$

$C = -\frac{1}{2}$

$\frac{-1}{2y} = ln(x-1) - \frac{1}{2}$

$y = \frac{-1}{2ln(x-1)-1}$

(c) Find $\frac{d^2y}{dx^2}$

$$\frac{d^2y}{dx^2} = \frac{\left[(x-1)\left(4y * \frac{dy}{dx}\right) - (2y^2)(1)\right]}{(x-1)^2}$$

$$\frac{d^2y}{dx^2} = \frac{\left[(x-1)\left(4y * \left(\frac{2y^2}{x-1}\right)\right) - (2y^2)(1)\right]}{(x-1)^2}$$

$$\frac{d^2y}{dx^2} = \frac{8y^3 - 2y^2}{(x-1)^2}$$

x	1	4	6	7
f(x)	10	8	5	2

5) Let g(x) be a twice-differentiable function defined by a differentiable function f, such that $g(x) = 2x + \int_1^{x^2} f(t)dt$. Selected values of f(x) are given in the table above.

(a) Use a Left Riemann sum using the subintervals indicated by the table to approximate g(3).

$$g(3) = 2(3) + \int_1^9 f(x)dx$$

$$g(3) \approx 2(3) + (3)(10) + (2)(8) + (1)(5)$$

$$g(3) \approx 57$$

(b) Find g'(3).

$$g'(x) = 2 + 2x \cdot f(x^2)$$

$$g'(3) = 2 + 2(3)f(9)$$

$$g'(3) = 14$$

(c) Using the data in the table, estimate f'(3).

$$f'(3) \approx \frac{f(4)-f(1)}{4-1} = \frac{8-10}{3} = \frac{-2}{3}$$

(d) Explain why there must be a value of c, on 1 < x < 7 such that $f(c) = 4$

Since g(x) is twice-differentiable, g'(x) is continuous therefore IVT applies. There must be a value of c, on 1 < x < 7, such that g'(c) = f(c) = 4 because f(4) > 4 > f(6).

6)

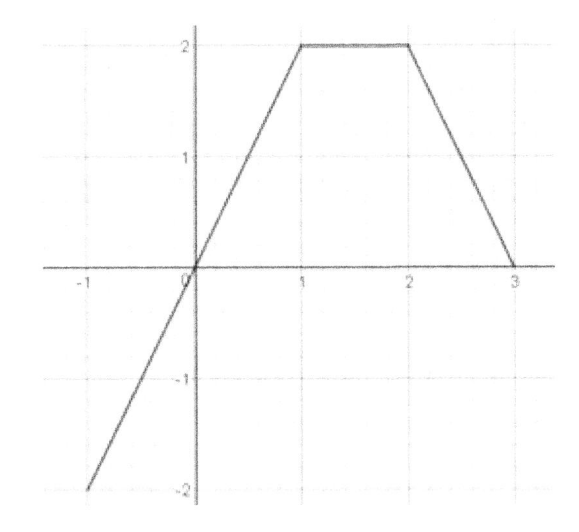

Graph of f(x)

x	g(x)	g'(x)
0	2	-4
1	1	-2
5/2	4	-3
5	-1	3

Let f be a continuous function defined on the interval [-1, 3], and whose graph is given above. Let g be a differentiable function with derivative g'. The table above gives the value of g(x) and g'(x) at selected x-values.

(a) Let h be the function defined by $h(x) = f(x) * g(x)$. Find $h'(0)$

$h'(x) = f'(x) * g(x) + f(x) * g'(x)$

$h'(0) = f'(0) * g(0) + f(0) * g'(0)$

$h'(0) = (2) * (2) + 0 * (-4)$

$h'(0) = 4$

(b) Let k be the function defined by $k(x) = g(f(x))$. Find the slope of the line tangent to k(x) at $x = \frac{5}{2}$.

$k'(x) = g'(f(x)) * f'(x))$

$k'\left(\frac{5}{2}\right) = g'\left(f\left(\frac{5}{2}\right)\right) * f'\left(\frac{5}{2}\right)$

$k'(\frac{5}{2}) = (-3) * (-2) = 6$

(c) Let $m(x) = \int_1^x g'(t)dt$. Find m(5) and m'(5)

$$m(5) = \int_1^5 g'(t)dt$$
$$m(5) = g(5) - g(1)$$
$$m(5) = (-1) - (1)$$
$$m(5) = -2$$

$$\frac{d}{dx}\left[m(x) = \int_1^x g'(t)dt\right]$$

$$m'(x) = g'(x)$$
$$m'(5) = g'(5)$$
$$m'(5) = 3$$

(d) Evaluate $\lim_{x \to 0} \frac{g(x)-2}{f(2x)}$

$\lim_{x \to 0} g(x) - 2 = 0 \; and \; \lim_{x \to 0} f(2x) = 0$

Therefore by L'Hopital's Rule

$$\lim_{x \to 0} \frac{g(x)-2}{f(2x)} = \lim_{x \to 0} \frac{g'(x)}{2f'(2x)} = \frac{g'(0)}{2f'(0)} = \frac{-4}{2*2} = -1$$

Explore Other books by Mirvoxid Press

Score Higher, Achieve More with Mirvoxid Press

these and more other Test Preps and Study Guides on Amazon Store:

Canada Store:

US Store:

Made in the USA
Monee, IL
09 December 2024

73078940R00059